Where Next, Columbus?

Where Next, Columbus?

New York Oxford 1994
OXFORD UNIVERSITY PRESS

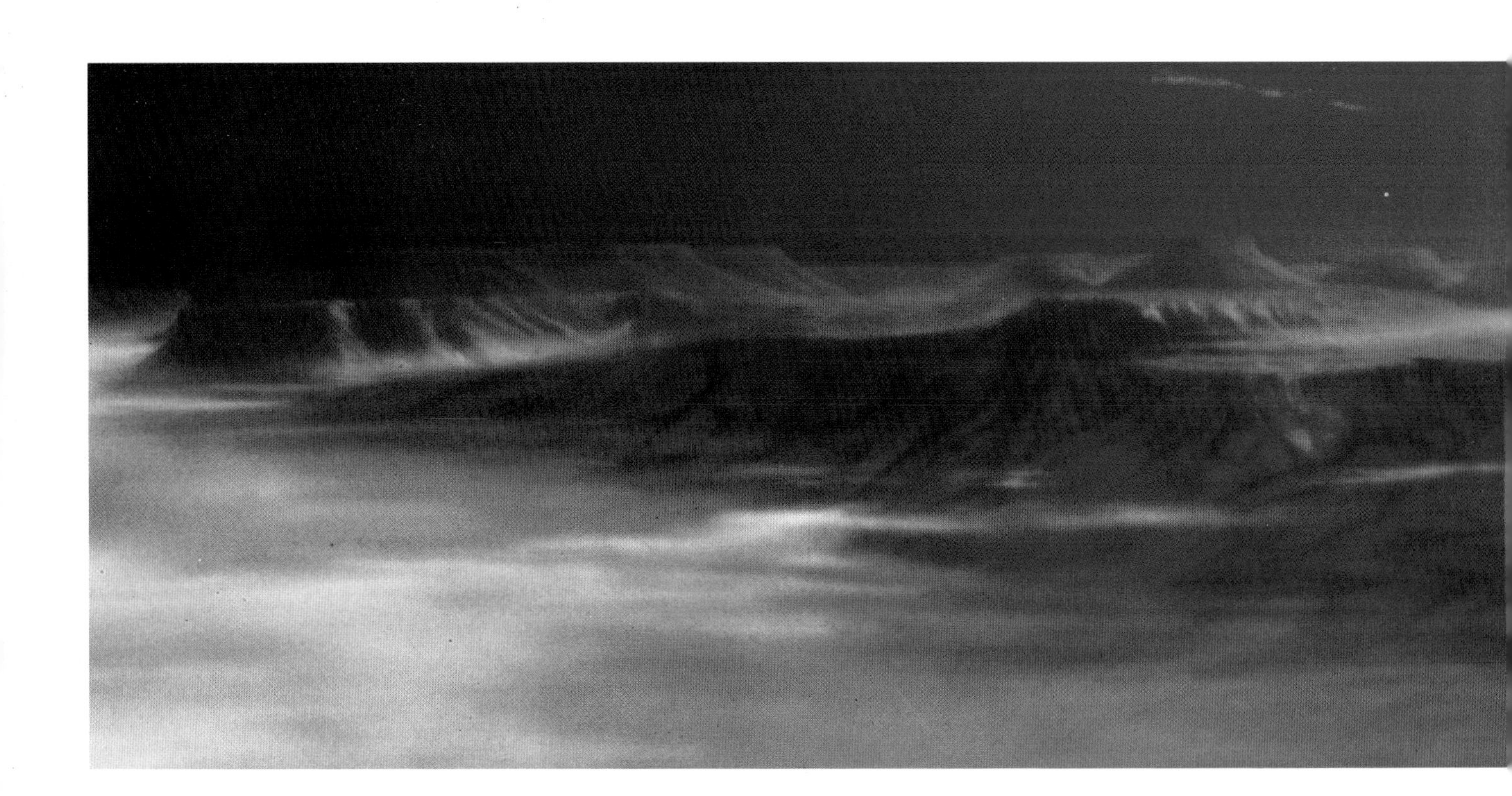

The Future of Space Exploration

Edited by
VALERIE NEAL
National Air and Space Museum, Smithsonian Institution

OXFORD UNIVERSITY PRESS

Oxford New York
Athens Auckland Bangkok Bombay
Calcutta Cape Town Dar es Salaam Delhi Florence
Hong Kong Istanbul Karachi Kuala Lumpur Madras
Madrid Melbourne Mexico City Nairobi Paris
Singapore Taipei Tokyo Toronto

and associated companies in
Berlin Ibadan

Published by Oxford University Press, Inc.,
200 Madison Avenue, New York, New York 10016

Oxford is a registered trademark of Oxford University Press

Library of Congress Cataloging-in-Publication Data
Where next, Columbus?: the future of space exploration/edited by Valerie Neal
p. cm.
ISBN 0-19-509277-5
1. Discoveries in geography. I. Neal, Valerie.
G80.W57 1995
919.904—dc20 94-11451

1 3 5 7 9 8 6 4 2

Printed in Hong Kong
on acid-free paper

Contents

Contributors

TIMOTHY FERRIS is an acclaimed science writer and a professor of journalism at the University of California, Berkeley. His essays have appeared in *Harper's*, *Life*, *Rolling Stone*, the *New Republic*, and the *New York Times*. His books include *The Mind's Sky*, *Coming of Age in the Milky Way*, *Galaxies*, and *The Red Limit*.

ROBERT L. FORWARD is a physicist and an aerospace consultant on advanced space propulsion. Formerly a scientist and manager with Hughes Aircraft Company and Hughes Research Laboratories, he has become a popular lecturer and author whose works include *Mirror Matter*, *Future Magic*, and the science fiction books *Dragon's Egg* and *Timemaster*.

STEPHEN JAY GOULD is a professor of geology at Harvard University, a curator of paleontology at the Agassiz Museum of Comparative Zoology, and a prolific writer. Among his prize-winning books are *The Panda's Thumb*, *The Mismeasure of Man*, and *Hen's Teeth and Horse's Toes*. He is active on the advisory boards of the Children's Television Workshop and the public television series "NOVA."

THOMAS E. LOVEJOY is a tropical and conservation biologist who has worked in the Amazon of Brazil since 1965. Before joining the Smithsonian Institution as Assistant Secretary for External Affairs, he held positions with the World Wildlife Fund–U.S. Lovejoy devised the innovative concept of debt-for-nature swaps, reducing international debt in exchange for conservation projects.

WALTER E. MASSEY served as director of the National Science Foundation from 1991 to 1993, after having been Vice President for Research at the University of Chicago and director of Argonne National Laboratory. A physicist with an active interest in science education, he is now Senior Vice President for Academic Affairs and Provost of the University of California system.

VALERIE NEAL is a curator in the Department of Space History at the Smithsonian Institution's National Air and Space Museum, and she is curator of the "Where Next, Columbus?" exhibition gallery there. She has written about *Spacelab* missions and orbital observatories for astrophysics, and she worked on the mission management support teams for four space shuttle missions.

STEPHEN J. PYNE is a historian whose special interests are science, exploration, and the American West. Now on the faculty of Arizona State University, he worked with the National Park Service as a firefighter and park ranger in the Grand Canyon and spent several months exploring Antarctica. His books include *Fire in America: A Cultural History* and *The Ice: A Journey to Antarctica.*

CARL SAGAN, a well-known author, astronomer, and ardent advocate for space exploration, is on the faculty of Cornell University and is president of the Planetary Society. Among his popular books are *The Dragons of Eden*, *Cosmos*, and *Contact.* Sagan's television production "Cosmos" stimulated widespread public interest in cosmology, space science, and exploration.

HARRISON H. "JACK" SCHMITT was the first scientist and the last astronaut to step on the moon. Lunar module pilot for the *Apollo 17* mission in 1972, Schmitt later served a term as U.S. senator from New Mexico. He now writes, lectures, and consults on a variety of issues in technology, energy, and space policy, including lunar bases and the exploration of Mars.

EDWARD C. STONE is director of the Jet Propulsion Laboratory and a vice president and professor of physics at the California Institute of Technology. Stone was project scientist for the grand tour of the outer planets by the *Voyager* spacecraft, and he developed instruments for research on more than a dozen other space missions. He lectures widely on space science and solar system exploration.

Where Next, Columbus?

Pat Rawlings, *Inevitable Descent* (1992)

VALERIE NEAL

Introduction

THIS BOOK has its origins in an exhibition at the National Air and Space Museum. For the quincentenary of Columbus's 1492 transatlantic voyage into the Western Hemisphere, the Smithsonian Institution sponsored a series of exhibitions, books, and symposia to analyze and interpret anew the significance of that event.

Most of the quincentennial programs were retrospective, with attention to the lingering legacies of the Columbian encounter in contemporary life. The National Air and Space Museum, however, decided to focus on the future in an exhibition called "Where Next, Columbus?" The purpose of this gallery is to stimulate thought about our prospects for future exploration and discovery of new worlds in space. It focuses on motives for modern space exploration, some of the options and choices in such an enterprise, and the challenges—physical, technical, and social—facing space explorers.

Exhibitions raise questions and present broad themes for consideration. But exhibitions are not books, and some ideas cannot be pursued as well on wall panels and in display cases as they can in essays. As curator of the "Where Next, Columbus?" exhibition, I envisioned a publication to complement the gallery, although not a traditional exhibition catalog to explicate the displayed objects in greater detail. Rather, I imagined a collection of thoughtful essays by people engaged in exploration or in imagining the future.

The writers whose essays are collected in this volume were invited to share their reflections on exploration. A student of Emerson and Thoreau, I savor the essay as an especially friendly and appealing literary genre. It is almost a conversation between writer and reader; together they roam through experiences and insights, make associations between events and ideas, and contemplate meaning. An essay is itself an act of exploration—an exploration of significant ideas and images.

The writers here share their thoughts about exploration past, present, and future. The collection opens with three essays about our legacy from past exploration. What has human experience meant? Historian Stephen J. Pyne surveys the past 500 years and distinguishes three eras, each characterized by a domain and style of exploration. *Apollo 17* astronaut Harrison H. Schmitt and Voyager missions' project scientist Edward C. Stone offer eyewitness accounts of two of the most successful expeditions of the present age; their observations and impressions are engaging.

The next four essays focus on exploration and scientific discovery. Ecologist Thomas E. Lovejoy urges a more environmentally aware technique of planetary exploration to avoid some of the mistakes made here on Earth. Paleontologist Stephen Jay Gould offers a compelling motive for close planetary scrutiny, particularly of Mars. Journalist Timothy Ferris argues for a change in attitude toward space exploration, while Walter E. Massey of the National Science Foundation looks at relationships between geographic and scientific exploration and discovery.

Finally, three essays examine our destiny in space and raise questions for the future. Astronomer Carl Sagan conveys a sense of urgency about the importance of continued exploration. Robert L. Forward, propulsion expert and science fiction writer, describes several ways to break out of the solar system and go to the stars. My essay probes the nature of commitment to space exploration.

The chapters of the book are bridged by brief commentaries on issues raised in the exhibition and the essays, and most of the illustrations are drawn from the "Where Next, Columbus?" gallery.

Decisions about the future have not yet been made. Recent studies and public discussions about the direction of the U.S. space program indicate that, a half century into the space age, the future of exploration is uncertain. This book is a contribution to continued dialogue on an intriguing question of our time: Where next?

LEGACY: PERSPECTIVES ON PAST EXPLORATION

HISTORIC events become meaningful as they are analyzed through lenses of scholarship and contemplation. The three essays in this section articulate dimensions of meaning in past exploration.

In the first, a historian examines circumstances, motives, and social forces to discern the significant relationships and trends that transcend particular events. Taking a broad perspective on the geography, psychology, and style of exploration during the past five centuries, he sees specific episodes as part of a set of patterns in history.

In the other two essays, a thoughtful participant offers the personal memories, emotions, and private moments that make up authentic human experience. Here, the reminiscence of one of only twelve men who lived briefly on the moon is juxtaposed with the recollections of a manager of robotic solar system exploration. Both accounts resonate with wonder.

Each commentary encourages us to consider the essence of the experience of exploration and the place of people in the enterprise. As an observer-historian, Stephen Pyne provocatively assesses the role of humans in space exploration, while Harrison Schmitt and Edward Stone convey a sense of immediacy plumbed from their own histories as explorers. Among the issues suggested by these essays is the importance of technology, motive, cost, and value. Is humanity's exploring role defined by the extent to which various technologies complement, exceed, or fall short of human abilities? What consequences and benefits are rewarding enough for societies to choose to continue exploring in space? Is it possible to distinguish between cost and value in decisions about human exploration of space?

These lingering questions from the past are germane to any future space-faring.

STEPHEN J. PYNE

Voyage of Discovery

PEOPLE are congenitally curious, and nothing is more natural than for them to work out that passion by moving around. The varieties of their travel are endless. Migration, walkabout, wanderlust, exile, war of conquest, trading expeditions, reconnaissance, long hunts, great treks, exodus, hegira and hajj, missionizing, pilgrimage, tourism, even espionage and enslavement—each is a kind of exploration, a way of learning about the world beyond one's immediate tribe or landscape.

Some of these travels, like the pilgrimage, are rituals, so formulaic in their movements and messages that they are cultural clichés. Others, like migrations and long hunts, allow for surprise. Some, a minority, really seek out the unknown. For Western civilization, a few modes of travel have assumed a special status complete with their own genres of literature, art, and scholarship. The quest, the grand tour, the journey to the underworld, the odyssey, and the trek beyond, for example, all propose lessons that transcend their historic times. To that august roster we should add the voyage of discovery.

The voyages of Renaissance Europe were logical successors to centuries of Western travel, trade, pilgrimage, conquest, and seafaring. The exploits of Alexander the Great, for example, may be thought of as a kind of exploration by conquest whose legacy included the fabled library at Alexandria. The pilgrims on the Camino de Santiago, Viking raiders to Greenland and Kiev, Crusaders to the Holy Land, traders to golden Guinea and the Great Khan, missionaries in search of Prester John in Asia and Ethiopia, peripatetic scholars trekking to universities in Paris and Oxford, shipborne colonizers to the Canaries and Madeira—all betrayed a restlessness that foreshadowed a more systematic pattern of exploration. The journeys of Columbus were part and parcel of a legacy of itinerancy, about to become a new species of travel—the voyage of discovery—that would shape how Western civilization learned about and assimilated the world around it.

The voyages of discovery catalyzed a process of world discovery, and once allied with the scientific revolution they resulted in a single world geography. They set in motion a dynamic of exploration—tied, on the one hand, to the geopolitics of European expansion, with its brawling rivalries, and, on the other, to the equally aggressive principles of modern science—that would prove irresistible. What Western explorers "discovered," of course, were places, peoples, and facts largely known to other peoples. But once those spheres of knowledge were assembled into a coherent whole, it was unnecessary for the process of geographic discovery to be repeated by other cultures. In a sense, there was only one world to discover. The voyage of discovery thus transcended its European origins and became a vehicle for global change. It has helped to remake the planet, and it has launched Spaceship Earth on its own uncertain voyage into the future.

Ideas, like travels, have their histories. "Voyage of discovery" has meant different things at different times, but among its critical attributes, I propose, are that it involves travel to some tangible place, that it fuses with the cultural paradigms that inform its age, and that it expresses the moral dilemmas engendered by encounters with new lands, peoples, and knowledge. Take away the geography, and the concept becomes only a loose metaphor that is useful poetically, but not very descriptive historically. Strip away the intellectual component, and the voyage becomes simply adventure or flight, miscellaneous wandering whose encounter with novelty, if any is found, is unexpected and unwanted. Pluck the voyage out of its moral universe, and it sheds its capacity to address fundamental questions of identity and purpose, to criticize or justify the society that launched it. Without that moral shock, the voyage becomes ritual or procession or the juvenile adventurism of pulp novels. When all three components are bonded together, however, as by a kind of cultural force field, the voyage of discovery becomes one of the most powerful expressions of its civilization—not only a probe that sends back data and images of the new, but a prism that refracts its sustaining society into its essences and transmits them back for self-examination.

Viewed in this way, it is possible to sketch three great eras of discovery. Each had its particular geography of exploration, each fused with different syndromes of thought and conventions of art and literature, each encountered

The grand gesture of the first great age of exploration was the voyage of circumnavigation. Magellan's expedition to sail around the Earth (1519–1521) epitomized Renaissance Europe's ventures of exploration and discovery.

John Carter Brown Library, Brown University

certain moral conundrums, and each advertised a distinctive expression of its ambitions, a kind of grand gesture. Each era, that is, was the creation of a particular civilization at a particular time in its history. The three eras were neither inevitable within the trajectory of Western history nor universal, although they may share universal longings and often tap archetypal motifs. Once established, the voyage of discovery underwent an evolution of its own, reforming to express the needs of its larger society. But its avatars have always retained a distinctive identity that has set the voyage of discovery apart from other travels and adventures.

THE FIRST great age of discovery began with the great voyages of Renaissance Europe. Its principal geographic accomplishment was the discovery of the world ocean; its distinctive expression, a voyage of circumnavigation. Of course, Europe explored more than the seven seas. Conquistadors, missionaries, and traders fanned out across the Americas, rode camels to the inland seas and bazaars of Central Asia, and even penetrated to Ethiopia and the fetid Amazon. With breathtaking speed, Russian Cossacks and *promyshleniki* swept across Siberia. But the new polities—the first truly global empires—required a maritime lifeline. Settlements tended to crowd the littorals of the world ocean, often perched like rookeries on offshore islands, reluctant to plunge inland other than to pillage or preach.

Gradually, as the magnitude of the discoveries became apparent, the influx of information, plunder, and experiences began to reshape European culture. Old paradigms of learning—the geography of the ancients, for example—crumbled before the onslaught. The New World could not be packaged within the inherited structures of the Old. The disparity between formal knowledge as espoused in the universities and the practical learning shipped back by voyagers became intolerable. It is no accident that the great voyages and the beginnings of the scientific revolution coincided. Francis Bacon thought that the printing press, the compass, and gunpowder had wrought more change than all the recovered texts cherished by Scholasticism; all three, not incidentally, were important for this sudden explosion in discovery—to disseminate its findings or to guide its ships or to render invincible the European vessels that sailed to foreign seas. When he wrote his *Instauratio Magna*, Bacon used for a frontispiece a ship sailing beyond the Pillars of Hercules, a symbolic voyage of discovery past the limits of ancient learning and geography. The Renaissance voyages set in motion the prospect for a global scholarship.

By sustaining new empires, the voyage of discovery also began to shape a global economy and a global ecology. Animals, plants, diseases, humans—all were shuttled around, sometimes deliberately, often inadvertently, as the camp followers of European exploration. Uninhabited islands became home to goats, rats, sheep, and people. Immigration, depopulation, and slavery rewrote the human geography of the planet. New foods and imported metals inspired a reformation of European agriculture and society. A global market sent tentacles to the farthest of lands. Sables in Siberia, beavers in the Rockies, rhinos on the African veld, sea otters along coastal California—all were exchanged for beads, manufactured goods, liquor, or gold. Political institutions adequate for relatively confined European fiefdoms broke down when stretched across oceans, and imperial accommodations in the theory and practice of governing affected Spain, Portugal, France, Holland, and England no less than their offshore trading factories, plantations, and colonies.

The voyage of discovery was inextricably bound up with this whole bubbling stew, a spoon that kept the pot astir. There was, initially, a vast curiosity to the age that absolves somewhat its swashbuckling savagery. But the voyages had particular purposes, usually commercial, either trade or plunder, although these ambitions were sometimes ameliorated by missionizing impulses. What the explorers learned, moreover, had a strongly ethnocentric bias. With few exceptions, they did not discover lands unknown to humanity, but only to Europe. Their maritime empires they did not invent out of whole cloth, but stitched together from patches plucked from the wreckage of those other maritime powers they encountered in the Pacific and Indian oceans, the Caribbean and South China seas. Discovery meant the transfer of knowledge from one society to another. Experienced Arab and Asian pilots guided European ships along the well-worn sea routes of the Indian Ocean and Indonesian archipelago. Interpreters instructed ignorant Europeans in proper behavior and survival skills. European explorers, and the savants and politicians who sought to assimilate their discoveries, found themselves engaged with peoples whose mores were fundamentally different from those of Christendom and for whom the Europeans had no suitable protocol of behavior, and who in fact often regarded the voyagers as unwashed barbarians and hucksters.

How Europeans, through their exploring emissaries, should respond provoked inquiries into political and legal philosophy. But the challenge went deeper. Western civilization began a great dialogue, ultimately moral, that questioned the foundations of its own ethical codes and broached the possibility of cultural relativitism. The myths of primitivism, the debates between

Exploration involved moral drama as well as geographic discovery. Both the exploring and the "discovered" cultures were challenged and changed by their encounters. New issues of power, wealth, belief, and behavior assaulted traditions at home and abroad.

Library of Congress

Bartolomé de Las Casas and Juan Ginés de Sepúlveda over whether the American Indian had a soul, the contrast between Black Legend and Noble Savage, between greedy invader interested only in plunder and power and (presumed) innocent indigene—the dialogue was not merely about the behavior of the new empires, but about their own character and legitimacy.

Exploration, that is, not merely projected European power outward, but also brought back counterexamples—of natives, heathen but benevolent; of outrages promulgated by European Christians—that questioned Europe's right to imperial expansion and that corroded, as by acid, the core of its ethical philosophy. Exploration involved moral drama as well as geographic discovery. It affected its sustaining civilizations as much as those they encountered. If discovery set off upheaval in overseas lands, so it inspired turmoil in Europe, whose folkways, beliefs, and institutions were eroded by the experiences of discovery and its returned plunder. The Spain that overthrew the Aztecs and Incas and that was soon awash in New World silver could not easily remain within the confines of Thomistic Scholasticism and medieval fealties: Indians were not Moors; the mythic and legal strictures of the Reconquista often broke when flung across the Atlantic or to remote Manila. The Britain that seized India, built railroads in Africa, erected penal colonies in Australia, and sent sledges to the South Pole was as much affected by those encounters as were the natives they met and the lands they inventoried; not

only did its economy and politics respond to these many varieties of exploration, but its view of the world, and its place in that world, necessarily changed.

Not surprisingly, perhaps, exploration entered into the creation myths of those new societies that Europe founded across the seas. The explorer typically enjoyed a more honorable status than the conquistador or the pirate or the acquisitive trader, even if, ultimately, his discoveries led to the same dark deeds. For new societies, the act of discovery testified to their honored origins as legitimate heirs, not bastard offcasts. They had been inspected, chosen, and recorded. The images that explorers brought back, or that intellectuals rewove from the fibers of their writings—visions of New Worlds, Virgin Forests, Noble Savages—served both to criticize the old ways and to promise that a better society was possible. They became literary and philosophical conventions, part of the moral canon of Western civilization. They deflated the pretenses of the parental society and celebrated the promise of the progeny.

For some time now, the dominant creation myth of America has followed just this formula: that America has resulted from the encounter of Old World civilization with New World wilderness and that from this sustained "discovery" derives America's greater purpose and moral vitality. If this myth is tragic as well as heroic, if the pioneer must destroy the circumstances that made him, the same can be said for the founding myths of Rome and Mesopotamia. What is special is the role of the explorer as the agent of national destiny, as the one who experiences that defining encounter for the first time.

Ethnocentricism was undeniable, saturating the concept of the voyage of discovery like a dye. If such voyages served as vehicles for self-examination, they also became occasions for self-congratulation. This was the one civilization that above all others explored. The true *mappa mundi* was the one it assembled and the others accepted. The drive to explore, it appeared, was built into the genetic code of Western civilization, which had to explore to be what it was. Without continuing discovery, the West would presumably suffer a crisis of courage, succumb to spiritual malaise, perhaps even wither into decadence and timidity. It needed the voyage of discovery, constantly renewed, to reaffirm its continued vigor and the rightness of its destiny. As the idea of progress seeped through its collective consciousness, intellectuals relocated a golden age of imagined virtue from a prelapsarian past to a futuristic utopia. The voyage of discovery became a vehicle for progress; the future belonged to those nations with the means and fortitude to explore.

This sense that discovery was destiny, that exploration was essential to the cultural metabolism of Western civilization, is well conveyed, if somewhat backhandedly, in that set piece of Western exploration art, the First Encounter. Romantics endowed the scene with wonder—the awe of discovering new lands and peoples, but also the awe with which the discoverers are viewed, as people of exceptional courage, fortitude, and curiosity. In fact, discovery was typically met with indifference or hostility, but artistic imagination spoke to self-perception, not to history. Each voyage of discovery reminded this civilization of what its peoples wanted to believe was best about themselves.

THE SECOND age elaborated these themes and altered them. Its geographic focus shifted from the littoral of the world ocean to the interior of the Earth's continents. This was, after all, the great era of European colonization and imperialism, an outmigration that swarmed over the Americas and Australia and that rewrote the political geography of Asia and Africa. The intellectual syndrome that shaped its curiosity was increasingly that of the Enlightenment, particularly the explosive imperialism of modern science. After a hiatus in which trade replaced discovery, national rivalries were renewed and the tem-

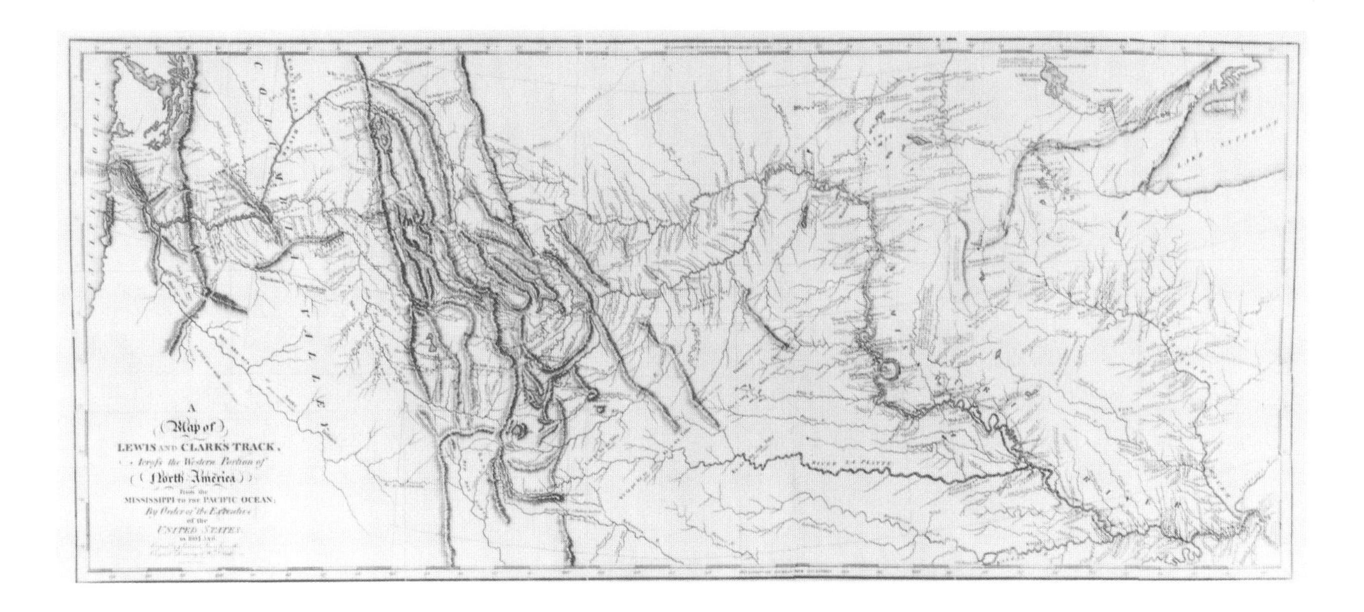

po of exploration quickened. The revival of circumnavigation in the mid-eighteenth century—the voyages of James Cook are exemplary—effectively announced the new age.

The grand gesture was the traverse of a continent, and its archetypal explorer was Alexander von Humboldt, the "second Columbus," the scientific discoverer of the New World and inventor of modern geography. As an explorer, Humboldt's achievement was to transfer the style of Cook to the continents. But in larger, cultural terms, he epitomized the explorer as Romantic hero, a man who stood to geographic discovery as Beethoven did to music and Napoleon to politics.

The continental traverses were the planetary flybys of their day, and they flooded existing institutions and theories with data and images. Information cascaded over all fields of cultural endeavor. Some new fields were watered and nourished; some old ones were washed away. The earth and life sciences assumed modern form and then unveiled new worlds, as the discovery of space nurtured the discovery of time: the far landscape of Earth history, with its vanquished geologic empires, lost worlds of fossils, and melancholy vastness. Painters rediscovered in nature a source of inspiration. Anthropology undertook a colossal inventory of the world's peoples. But the age's most enduring symbol was, of course, Charles Darwin's exposition of evolution by natural selection, a by-product of his own 5-year voyage of discovery on the HMS *Beagle*.

The characteristic expedition of the second age of exploration was the trek across a continent. The Lewis and Clark expedition across western North America (1804–1806) exemplified Enlightenment-era expansionist and scientific impulses.

Library of Congress

The moral dilemmas of discovery, too, were secularized. The expansion of Europe acquired a kind of inevitability, a destiny to bring to savage and unenlightened peoples such benefits of civilization as democracy, science, Christianity, French, and long pants. What in the United States became known as manifest destiny had its echoes around the globe as Europeans sought to explain and justify their aggressive penetration into every valley and their pursuit of ever more remote and elusive tribes.

But inevitably, as explorers assumed the dress, spoke the language, and practiced the customs of their guides, hunters, and bearers, they experienced new values and saw the world in new ways that challenged the comfortable clichés of Western superiority, that doubted whether technological progress also meant spiritual progress, that questioned whether political ascendancy necessarily meant ethical superiority. Instead, those experiences often suggested an equivalence of moral worlds. American literature of the nineteenth century repeatedly equipped its adventuring hero with a dark-skinned companion of equal virtue, if lesser social standing. The Eskimo hunter, the Indian scout, the South Seas cannibal qua whaler—all questioned the presumed primacy of Western values, as Europeans decimated the flora, fauna, and indigenes of whatever lands they touched. A new literary convention entered Western consciousness—the Last of the Mohicans, the final, poignant specimen of a near-vanquished breed, be it tree, buffalo, or tribesman. Pioneering, it was real-

As these two contrasting perspectives suggest, the presence of resident populations was a salient fact in previous ages of exploration, shaping and also raising doubts about progress and destiny.

Above: Charles M. Russell, *Indians Discovering Lewis and Clark* (Montana Historical Society, Mackay Collection); *below:* Olaf Seltzer, *Lewis and Clark at the Great Falls of the Missouri* (Thomas Gilcrease Institute of American History and Art)

ized, was a tragic act that destroyed the very conditions that made pioneering possible. Colonization could be ironic, often replacing the noble with the sordid. The voyage of discovery was not necessarily the bold first strike of enlightenment, but might announce the toxic traces of an approaching moral pollution that foreshadowed plunder, disease, and dissolution.

Such doubts came slowly, however. Exploration stalled primarily for other reasons. The last of the continents, Antarctica, resisted discovery by traditional means, a sink rather than a source. The heroic age of Antarctic exploration proceeded in defiance of the natural order of geophysics. The explorer was less the rational emissary of science and empire than an Ahab maddened by his pursuit of the white whale. The ice field that composed Antarctica steadily stripped away the kind of data, experiences, and referents that discovery had become accustomed to; the continent was ringed by steepening gradients of energy and information, rich on the outside, barren toward the center; a journey to the source meant that there was more and more of less and less, until in the end there was only ice and self. Expeditions scrounged for survival. Dialogues with the Earth became soliloquies with the self. There were no native peoples to serve as guides, no alternative moral order with which to contend. There was only ice. The reductionist ice became a terrible mirror that first stripped the viewer and then reflected back what remained. On the interior ice sheets, the traditional

The last of the continents, Antarctica, resisted discovery. The ice field lacked the kind of data, experiences, and referents that had characterized other explorations.

Stephen J. Pyne

mechanisms of exploration went blank. Technology and culture could send explorers to Antarctica, but they could not outfit them to understand what they found.

Then came the Great War. In its aftermath, the militant enthusiasm for imperial adventure cooled. The need for boundary surveys across Africa, for geologic prospecting in Asia, and for competitive scrambles across the Antarctic plateau vanished, almost overnight. Meanwhile, modernism pummeled the cozy rationalism of the Greater Enlightenment. The second age limped off, mortally wounded, to the rain forests of New Guinea, remote tribes in the Red Centre of Australia, and isolated mesas within the Grand Canyon that were rumored, like some lost world, to shelter weirdly evolved species.

The second age lost its distinctive geographic dominion, its intellectual sustenance, and its moral imperative. For a civilization convulsed by political and intellectual revolutions—by two world wars, a global depression, and totalitarian dictatorships of the right and left; by an ontological reformation dramatized by relativity and quantum physics and by biology's "new synthesis"; by the shock of cubism and dadaism and their proliferating offspring that overthrew the premises of Western art—the voyage of discovery seemed as archaic as a caravel. It could interest only dilettantes and antiquarian adventurers—the Buffalo Bills of discovery—who did not know that the frontier had closed and that the only important discoveries left must come from laboratories and

the self-reflective minds of artists. The celebrity replaced the hero. The Romantic explorer had become himself a Last Mohican.

THE THIRD great age of discovery began, as had the two before it, in the oceans. Spurred by a world war, submarines and remote-sensing devices steadily mapped the contours of the ocean floor, roughly 75 percent of the Earth's solid surface. Interest revived also in the polar regions: the Arctic because of its strategic importance in the emerging cold war; the Antarctic, a *terra nullius*, because it could now be surveyed, even its sub-ice terrains, with the logistical power and new instruments developed during the war. A proposed Third Polar Year quickly escalated into a more ambitious global inventory for which primitive spacecraft could also be mobilized, what became the International Geophysical Year (1957–1958). IGY effectively announced the third age.

The third age took for its geographic domain the solar system, and for its grand gesture, the geophysical inventory of a planet. Planet Earth inaugurated the era, much as the first age had nurtured its maritime technologies by plying European seas and the second age had field-tested itself with a resurvey of the European land mass. The deep oceans, with their trenches, globe-encircling ridges, guyots, and abyssal plains; near space, with its stratospheric winds, radiation belts, and exotic auroras; glacial-clad Greenland and Antarctica, with an ice sheet so immense it deformed the planet—all constituted important environments. Those territories, never before occupied by humans, stimulated new theories of geopolitics and novel institutions by which to administer them. The belief that sovereignty had to reside in some state had little meaning where permanent residence was impossible, where environments repelled human life of any kind. *Terra nullius*, the land of no one, became *terra communis*, the land of everyone. The Antarctic Treaty, which entered into force in 1961; the revised Law of the Sea, which provided for the potential exploitation of the sea floors; and the Outer Space Treaty (1967), upgraded into the Moon Treaty (1979)—each attempted to incorporate into international law the bold new landscapes made accessible by the third age, and each, in its own way, proposed that such environments were the "common heritage of all mankind."

These accommodations, however, are symbolic of a larger cultural reformation. Decade by decade, modernism had reworked the intellectual legacy of the Greater Enlightenment, its plastic arts, its literary genres, its sciences. The absolute had become the relative; the perspective of the omniscient narrator, the voice of irony; the march of progress, the descent into existentialism. The great breakthroughs in physics and mathematics had not, as had the scientific revolution in the sixteenth and seventeenth centuries, required the stimulus of an expanding world. The modern synthesis in the life sciences had not, as had the Darwinian synthesis in the nineteenth century, built on the discoveries of a great age of exploration. Too long lashed to an antiquated source of data, like partners in a three-legged race, the earth sciences lagged badly, conscious of a missed revolution. What had given character to past ages was not geographic discovery alone or speculative contemplation, but their interaction, like the two poles of a magnet that defined between them a specific force field. Take away one, and the field collapses.

By the time IGY bundled its ice cores and published its figures for solar flares, however, modernism had invaded intellectual culture like a computer virus and began rewriting the software of Western civilization. The third age sent modernism into space. The alliance, ever awkward, was essential. Without

The domain of the third age of exploration is the solar system, and the principal gesture is the geophysical inventory of each planet. As the frontier extends into space, strange new landscapes come into view. These composite images are the handiwork of modern explorers' instruments, spacecraft, and computers.

Below: surface of Venus (NASA); *right:* photomosaic of Mars (U.S. Geological Survey)

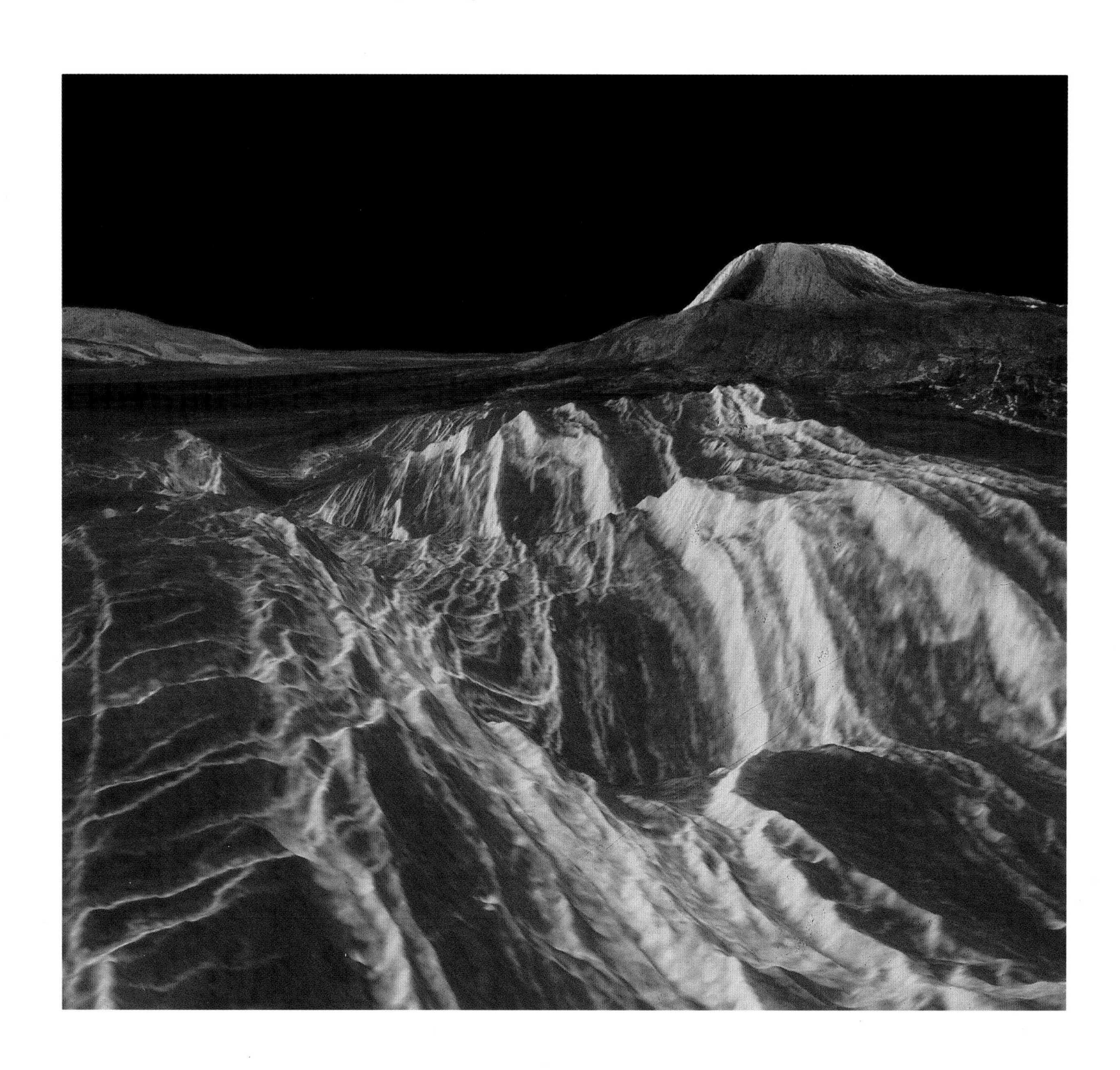

VIKING
MARS
REV 334 - 6
STRETCH RED: 0- 0; 30- 0; 180-255; 255-255
STRETCH GREEN: 0- 0; 30- 0; 230-255; 255-255
STRETCH BLUE: 0- 0; 30- 0; 245-255; 255-255
POSITIVE 25RASTER 8BIT 6250BPI
USGS

that cultural connection, exploration lacked intellectual vitality, and all its splendid technology would add up to nothing more than military adventurism or the intellectual equivalent of stock-car racing, a form of communal recreation rather than of collective curiosity. Equally, without some vehicle to force it to look outward, to confront phenomena outside itself, modernism threatened to spin steadily into a black hole of solipsism.

Now the technology that took Western civilization into space had the means by which to understand what it found. The landscapes of the third age—abstract, conceptual, minimal—were the work of nature as modernist and could be appreciated accordingly. The scientific data that poured in swept over intellectual paradigms like an avalanche. Some new planetary sciences appeared; some old ones were reformed (geology most notably, the theory of plate tectonics emerging more or less directly from the renewal of exploration); and some, like anthropology—too keenly dependent on the continual discovery of new peoples—silently declared the intellectual equivalent of Chapter 11 bankruptcy. There was no anthropology of the solar system.

This fact highlights one of the vital differences between the third age and its predecessors. A few oddities of life could be found in its terrestrial landscapes: lichens on exposed Antarctic sandstones, wormlike creatures attached to oceanic hotvents. But apart from the Earth, there was nothing. No life. No peoples. No intelligence. Whatever of these might be found outside the solar system placed their discovery beyond the pale of the third age. The third great age of discovery, our age, would have to achieve understanding without interpreters, translators, native guides, hunters, and collectors. No one could live off the land, go native, absorb the art and mythology of an alien consciousness, experience an alternative moral realm. There would be no one to talk to except ourselves; despite, or perhaps because of, attempts to treat it according to the prescriptions of the past, discovery inexorably becomes a colossal exercise in self-reference and self-reflection. Beyond the Earth there would be no morality, as traditionally understood. The special geography of the third age leaped beyond the practical bioethics of even deep ecology.

Accept this premise and some interesting corollaries follow. The third age could avoid the moral dilemma of earlier imperialism in which the expansion of one group came at the expense of another. There is only one culture, that of the civilization that launched the voyages of discovery. Ethnocentricity vanishes. There is no exo-ecosystem to foul, only the worry that, in some unexpected fashion, contaminants might be brought back that would harm the Earth's own biosphere. With no distinctively *human* encounter possible, there is no compelling reason for humans to even serve as explorers. Robots and remote-sensing devices can gather the critical information. Cameras can outfit everyone having access to a television set with his or her own personal "eyes of discovery." It is possible to retrofit, without undue moral qualms, high-tech hardware

onto the jingoism of earlier expansionism. Equip manifest destiny with warp drive and transporters. Locate the New Jerusalem on Mars. Listen for Prester John among distant galaxies. Carry the white man's burden to Saturn. As long as other life or cultures are absent, there is no moral crisis except whatever we choose to impose on ourselves. The morality at issue is one of the self, not between the self and an other.

But if the third age avoids some of the problems that compromised and made moribund the second age, it poses its own peculiar dilemmas. Where is the compelling *human* interest in such discovery? Where is its excitement, the passion that ultimately comes from moral drama? How can great literature and art come from such nonencounters? From where comes the requisite public enthusiasm to sustain costly ventures?

In the present age, exploration no longer has *to be done by on-site people; robots and remote sensors like the first visitors to Mars—the* Mariner 4 *and* Viking *spacecraft—can gather ample information as surrogate explorers.*
NASA

That the technology exists to retrofit the voyage of discovery into the solar system does not mean that such voyages will, or should, be launched. That intellectuals are curious about the new discoveries, or that certain arts and sciences may benefit from them, does not mean that a society will elect to pursue them. Society must have a reason to choose such discovery, to believe in it, to thrill or despair to its revelations; the justification for exploration must transcend the political rivalry of a defunct cold war; people must find a greater meaning in planetary discovery than as a national hobby or a jobs program or a kind of long-running daytime television. To fail is to agree with a mordant Kurt Vonnegut in *The Sirens of Titan*, written as IGY concluded:

> The state of mind on Earth with regard to space exploration was much like the state of mind in Europe with regard to exploration of the Atlantic before Christopher Columbus set out.
>
> There were these important differences, however: the monsters between space explorers and their goals were not imaginary, but numerous, hideous, various and uniformly cataclysmic; the cost of even a small expedition was enough to ruin most nations; and it was a virtual certainty that no expedition could increase the wealth of its sponsors.
>
> In short, on the basis of horse sense and the best scientific information, there was nothing good to be said for the exploration of space.*

* Kurt Vonnegut, Jr., *The Sirens of Titan* (New York: Dell, 1970), 30.

The dangers were indeed great, the costs huge, and the economic rewards uncertain. After the collapse of the cold war, the geopolitical pressures quickly deflated. But there was much good to be said for the enterprise, and there is no valid reason to think that reprovisioned voyages of discovery cannot capture the sentiments of their predecessors, including their ties to literature. The exploration of Antarctica showed how an art and a literature could develop, even in the absence of a robust modernism. (Besides, if postmodernism can survive deconstructionism and the posturing of structuralist architecture, it can cope with the intellectual challenges of the third age.) Well before the third age, science fiction had rushed in to populate the void of interplanetary space with an imagined geography, an epic history, even a folklore. Once the age matured, the genre enjoyed an impressive revival: "Star Trek" renewed the voyage of the *Beagle* at warp drive; "Star Wars" retrofited *Ivanhoe* with light sabers. If in real life the *Voyager* spacecraft found no other space-farers, they helped inspire a second-order literature. The absence of moral drama of the old kind is more than compensated by the absence of moral conflict of the old kind. The loss of received awe—the fact that there is no one to experience wonder at the arrived explorers and their civilization—may be happily discarded as excess cultural baggage, overload for a people already steeped in narcissism.

Perhaps the central conundrum is the role of humans. The exploration of

In space thus far, explorers have found only barren, inhospitable worlds devoid of life. Whereas in previous eras, exploration involved the moral drama of encounter, modern exploration is marked by solitude.

Overleaf: NASA

the solar system would encounter no other people, and this fact made unessential the use of human explorers. In past ages, discovery *had* to be done by people. There was no other option by which to learn the language, to record data and impressions, to meet other societies and translate their accumulated wisdom. It is impossible to imagine the great expeditions of the past without considering the personality of individual explorers who inspired, directed, witnessed, fought, wrote, drew, exulted, feared, suffered, and otherwise served as antennas and lightning rods for their civilization.

But humans do not have to be physically present at the discoveries of the third age, and there are good reasons for arguing that they should not be. Robots, cameras, and high-tech instruments can get to the critical environments, record the sights, and take the necessary measurements. With television, the revelations can be broadcast instantaneously and equally for all to witness. Moreover, robotic exploration can do this for far less cost and less risk than required by manned voyages. What such spacecraft cannot do is record the experience of actually being there. But what does that mean, when such hostile environments deny one the chance to touch, smell, taste, or communicate directly with one's surroundings, when an explorer can survive only within a completely artificial environment? The one immediate sensation is vision, which television can disseminate readily across interplanetary space. Voyages of discovery need no longer sail with their perceptual rudder unique-

ly bound to the hand of an individual explorer. And they need no longer put human life at risk, not in the way that, in the first age, wooden ships might flounder in Caribbean hurricanes or, in the second age, exploring parties might be attacked by desert tribesmen or fire-wielding Aborigines.

I BELIEVE that the voyage of discovery is indeed one of the great inventions, even an attribute, of Western civilization and that its loss would pauperize our civilization of useful data and a vital spirit. Something would go out of the United States in particular—"exploration's nation," as William Goetzmann has called it—if the tradition of geographic discovery disappears. But it is a mistake to assume that the voyage of discovery is itself unchanged or must remain unchangeable if it is to survive, that space exploration must proceed in the same ways as the voyages of Columbus, Vasco da Gama, and Magellan or the cross-continental expeditions of Humboldt, Peter Pallas, and Lewis and Clark. Of course, the need for human commentators, scholars, artists, and scientists remains, but not for human explorers, not as traditionally understood. The inhuman quality of the third age can, in reality, become an asset.

The voyage of discovery is an invention like light bulbs, landscape painting, and political constitutions. It was created at a particular place and time, shaped by the materials at hand; it has evolved, often with culturally biased characteristics and under the prod of national rivalries; and it will continue to metamorphose if it is allowed to adjust to changing circumstances of geography, intellectual syndromes, and moral concerns. It has to change. The United States survived another nominal crisis of its geography, the "closing" of the frontier, by relocating the desired virtues elsewhere (endowing the hard-boiled detective with the literary character of the frontiersman, for example) and by reserving public lands (setting aside wildlife reserves and forests where big-game hunting could continue, and creating parks where the sense of wonder, of personal discovery of the wild, can be reenacted). If a nation can do this, it can accommodate the changes demanded by the third age of discovery.

The precedent exists. In his poem "Ulysses," Alfred Tennyson reworked Homer's *Odyssey* into a kind of credo for the Romantic explorer, transfiguring a beleaguered wandering into an act of exploration, reinvigorating a bored Odysseus with a challenge to sail beyond the sunset. In a nutshell, that is what Western civilization did when it invented the voyage of discovery. But that particular credo ended on the Antarctic ice, and its admonition "to strive, to seek, to find, and not to yield" is appropriately carved onto the memorial cross dedicated to the British explorers who reached the South Pole and failed to return. They and the tradition they represented had gone beyond the bounds of the second age.

What is needed now is another such transfiguration. What is needed is a full-blown third age. The technology exists; the intellectual syndrome by which to interpret it is at hand; and the requisite moral drama, if not wholly resolved, is at least stripped of its worst ambiguities. Allow the exploring tradition to undergo this transformation, and the voyage of discovery will carry us into the twenty-second century and sail beyond the solar system.

HARRISON H. SCHMITT

A Trip to the Moon

A LEAVE TAKEN

TUBULAR beams from the searchlights turned our Saturn V rocket into a gleaming white spike that pierced the night toward a thin crescent moon. Reflected spears of light played on lines of low offshore clouds and, in anticipation, sped between those clouds through misty air into space.

The night before the launch of the last *Apollo* mission to the moon, I viewed, again, this scene from space age memories. I had driven toward Florida's Atlantic coast and Pad 39, past the launch-control block house, the surreal and huge Vehicle Assembly Building, and the giant silhouetted "crawler" that had carried the rocket on the first leg of its journey many weeks before. A few days before Christmas just 4 years earlier than this December night in 1972, I had joined another rookie astronaut, *Apollo 8*'s Bill Anders, on a nighttime visit to Pad 39. The Gothic analogies of the illuminated Saturn first came to mind that night—a brilliant white spire, built by the dedication of so many, projecting its own architectural message of the future into space.

On December 6, 1972, however, the Saturn V seemed to have grown even larger and higher as Gene Cernan, Ron Evans, and I walked single file, 315 feet above the launch pad, across a catwalk's open mesh floor. As we looked down, the familiar statistics—364 feet high, 6.2 million pounds, more than 1 million components—became meaningless. The huge white cylinders of the three Saturn stages dominated everything. Streamers of water vapor,

produced by venting from tanks of cold fuel and oxygen, swirled around us and the rocket. Through the mists flashed the words "United States"; the red, white, and blue flags; and the red-lettered "USA," painted so precisely on the rocket's gleaming sides.

Waiting patiently, the lunar module *Challenger* rested folded and unseen atop the rocket's third stage, positioned just below the command module *America*. Outside and below *Challenger*'s compartment, the crusted icy condensation glistened on the bright white skin of the liquid-oxygen and -hydrogen tanks. In the swirling vapors and arc-light glare, the world of family, friends, and country disappeared from view and thought.

The protective "white room" surrounding the open hatch of *America* provided our last direct contact with colleagues who helped us through the final preparations. With communications and environmental-control checks complete, we closed the hatch and locked, checked, and double-checked the

The Apollo 17 *crew were commander Eugene A. Cernan (seated), command module pilot Ronald E. Evans (right), and the author, lunar module pilot Harrison H. Schmitt.*
NASA

emergency egress system. The tragedy of the fire that had destroyed *Apollo 1* and killed its three-man crew in January 1967 still lurked just below the surface of our thoughts as the catwalk arm and attached white room moved away to a temporary position, ready to return quickly.

On the form-fitted couches inside *America*, the three of us now lay both alone and not alone. Half a million friends had put us there and still watched over us—the planners, the designers, the builders, the test teams, the trainers, the support teams, and, finally, the launch and flight controllers. The tangible and intangible results of their spirit and ingenuity enclosed and sustained us. Thousands would see their jobs eliminated as we lifted off for the moon, but not one failed us. Few, if any, regretted the sacrifices that they and their families had made on our behalf and on behalf of their country. At this moment, each believed, truly believed, that the Apollo program had become and would remain the most important commitment they would make with their lives!

For the present, all of this lay outside the conical world of *America*. The years of thought and training and planning and preparation had placed us in a quiet capsule just above the lunar module *Challenger*, one-fifteenth of a mile above the Saturn's waiting first-stage engines.

"Thirty seconds and counting." The quiet broke as the rocket awakened. Like an immense cold-blooded animal, moving, stretching, and alive beneath us, the rocket aligned its great engines, pressurized its web of fuel and oxidizer lines, and checked its electronic vital signs one more time. Somewhere, using millions of bits of information coming in over thousands of wires, the launch computer complex also performed the last series of checks, as it had been programmed to do long ago by persons unknown to us. Faster than an eye blink, the launch computer did what years of analysis and thought had designed it to do. It stopped the launch countdown!

"We have a hold." Instantly, the calmly spoken words came to us. Then, "30 seconds and holding."

As the spacecraft's launch timer counted down from 30 seconds through zero, we lay there without moving, three pairs of eyes rapidly scanning the instrument panels for any clues to why the countdown had stopped at this critical point. "Nothing wrong on my gauges." Did the Saturn V know that we were at "30 seconds and holding"? We felt the rush of adrenalin and waves of disappointment; then anticipation of the violence of a possible launch-pad abort tensed every nerve and muscle.

After about 5 minutes of extreme uncertainty mixed with not-so-reassuring reports on the communications loop that "we're looking at the problem," the calm and calming voice of Skip Chauvin, the launch director, entered our ears: "It looks like the launch computer thought the LOX [liquid oxygen] tank for the S-IC [first stage] wasn't pressurized even though we had pressurized it with a manual command from here. It's either a sensor or a computer problem. The Saturn's fine. We'll work it out and get back to you shortly. By the way, we've recycled to T minus 8 minutes and holding."*

His voice carried a tone of competence and confidence that immediately told me that "we know what the problem is, we know how to fix it, and all we have to do is convince the bosses we can go ahead." With this feeling of assurance and the soft hum of the radios in my head, I went to sleep. Nothing I could do would help, and you never know when you may need the rest.

More than 2 hours after our first experience with "30 seconds and counting," everyone who needed to be satisfied had been satisfied with the planned "fix." A new wire placed in the reluctant launch computer "told" the computer to ignore the error in its program, and, again, we began at "T minus 8 minutes and counting." The countdown toward our future had restarted, this time for good.

Half a minute to go again! This time, the preparatory antics and murmurs of the Saturn seemed familiar. At T minus 4 seconds, heavy massaging vibrations began against our backs as the five F-l engines started and quickly reached their full power. Thousands of gallons of kerosene and liquid oxygen fed an insatiable appetite as the huge hold-down arms held us to the Earth for

* NASA transcript of the *Apollo 17* mission.

The final U.S. mission to the moon, Apollo 17, *was launched on a trail of brilliant flame, splitting open the night.*

NASA

the last few seconds before releasing us to space. In one more moment, 7.5 million pounds of thrust would lift us toward orbit on a trail of brilliant flame, splitting open the night.

Pulsing waves of sound and flashing streams of light buffeted the bodies and minds of onlookers, bringing spontaneous and unexpected hugs, cries, and tears. Once again, a life-tipped pillar of fire, the Saturn rocket, a massive tribute to boldness and imagination, became a blazing symbol of human potential for greatness.

"We have liftoff!" Forces of acceleration and vibration gradually increased, adding to the new experiences. Two minutes after liftoff, amid intense shaking and noise, our weight reached four times normal, and I wondered about having the strength to reach switches in the event of malfunctioning spacecraft systems. The instrument panel oscillated so fast that most of the dials and switch labels had become unreadable.

Suddenly, the first-stage center engine stopped. Had 2 minutes and 20 seconds gone by already? Then, 25 seconds later, right on the money, the four outer engines stopped, the first stage separated, the five second-stage engines ignited, the skirt between the first two stages tumbled away, and we continued on our way once again. This sequence took us from a positive acceleration of four times Earth's gravity (g), through a negative acceleration of 1.5 g's as the entire rocket unloaded under no thrust, and, finally, to a positive acceleration of 1.5 g's on the second stage. All of this happened in just over a second! That got my attention!

Being propelled by the five hydrogen–oxygen engines of the S-II (second stage) felt like a ride over deep velvet by comparison with the raw violence of the first stage. Other familiar milestones passed: "launch escape tower jettison," "maximum aerodynamic pressure," and "third-stage S-IVB to orbit capability." At 9 minutes and 20 seconds, the shutdown and staging of the S-II and the ignition of the S-IVB came as planned.

"Mark. Mode IV capability." Nine minutes and 23 seconds, and, from the capsule communicator (CAPCOM) report, we knew that we could at least get into Earth orbit on *America*'s single rocket engine if the S-IVB quit on us. Any concerns were unwarranted as the old S-IVB purred away with every indication that it would serve us well in the long acceleration from 18,000 miles per hour in Earth orbit to the nearly 25,000 miles per hour needed to escape Earth's gravity and make "a trip to the moon."

A FIRST LOOK

"SEVENTEEN, Houston. You are GO for orbit." Insertion into Earth orbit approached with increasing anticipation. For Ron and me, this would be our first real exposure to space! We had earned our gold astronaut pins several minutes earlier as an altitude of 50 miles went past;

however, true space flight would be very different. Even though we watched the computer displays and mission timer count us down to 11 minutes and 42 seconds and the attainment of a nearly perfect 90-nautical-mile circular orbit, the shutdown of the S-IVB and sudden weightlessness came as a complete surprise. No weight! Weightlessness no longer existed as a fleeting feeling from a childhood carnival ride or the few tens of seconds on parabolic training flights in a KC-135 transport. Now, the real sensation went on and on!

"Hey, where did that panel screw come from?" Every object not fastened down or stowed floated by, headed eventually for the screen on the cabin intake fan. Moving from place to place felt like swimming without any water. How could anyone have ever called it space walking? Only the caution learned by listening to others who had been there before and the demands of the checklist kept me from playing with the abandon of a child.

Soon, the windows and the first view of Earth from this orbital perspective beckoned in spite of the checklist demands and some physical uneasiness. Like our childhood home, we really see the Earth only as we prepare to leave. The years of flying T-38s at 40,000 feet above most of Earth's weather have not prepared us for orbiting the planet at 540,000 feet. New patterns suddenly jump into view. Over the South Atlantic, between Africa and South America, a solid layer of morning clouds lies crenelated from one horizon to the other like an old washboard. Jagged breaks in these wrinkled stratus give the illusion of polar pack ice fracturing and moving apart.

The immensity of continental Africa next holds the eye. The first view of land includes the giant sand dunes on the northwestern coast. A pass over cloud-free Africa and Madagascar at midday offers aesthetic and geologic scenery unlike that in any textbook or travelogue. Geology and botany define ageless, still mysterious patterns of life and soil that, millions of years ago, nurtured our ancient ancestors. Preceded by long lines of billowing white smoke, range fires sweep before the winds. The fires emblaze grasslands with feathery patterns like ice crystals on a window after winter's breath has passed.

As we speed silently over the deep blue Indian Ocean, dense white tops of hundreds of magnificent afternoon build-ups break into the stratosphere, forming where tropical air masses converge near the equator. Feeding moisture into air currents that ultimately drop snow in the polar regions, these splendid examples of nature's cycles and nature's might dominate our paths over equatorial regions. The sunset shadows of these thunderheads stretch westward like dark fingers reaching to stop the inevitable advance of night. Approaching the Indies, the flashing blue-white lightning in these storms ripples through the clouds and across the ocean's black night.

North and south of our orbit, stars rise slowly between the dark horizon of the night and the soft blue airglow in the upper atmosphere. I tell friends on the airwaves that "a field of stars on the Earth is competing with the heavens. . . . [We] are going right over Florida now, looking down at Miami. A beautiful view of the Keys all lit up . . . , and I just saw a shooting star right over Miami!"*

Banded sunrises change in a few minutes from deepest black to brilliant blue to desert orange to purest yellow to searing daylight and then back to sunset in never-ending progression. The Earth displays tinted oceans and quilted continents with patterns wrought during 4.5 billion years, dark-shadowed thunderheads and dazzling snows ever varying in their mysteries and beauty, and civilization's warm fields of lights by night and farms by day, seen without the political and racial boundaries we learned in other times.

The urgency of preparations for leaving these fascinating scenes of Earth intervened until I was able to steal a nighttime glance at a cloud-covered New Mexico and home. Only diffuse glows of the lights of Albuquerque and El Paso penetrated late autumn clouds. Florida appeared again. We knew it was clear at Cape Canaveral because we had left only one orbit earlier, just an hour and a half ago. An hour and a half! It hardly seemed real to the space rookies, no matter how much we had heard before.

---

* NASA transcript of the *Apollo 17* mission.

"Seventeen, you're looking great on the final status check here, and you're GO for TLI [translunar injection]." The last hurdle before coasting into deep space and a rendezvous with the dark leading edge of the moon lay before us. "The light's on and we have ignition," Gene reported. "Seventeen, Houston. You're looking good, and the thrust is GO."* For 5 minutes and 15 seconds, we accelerated ahead of the Saturn V's soundless third-stage rocket, fascinated both by the marvel of the human and technological event of which we were a part and by the unbelievable beauty of the ribboned sunrise through which we sped.

A perfect burn, and *Apollo 17* had truly uncoupled the gravitational chains of Earth. Once again, humans had left their evolutionary cradle for another potential home in space.

A NEW VIEW

THE LIFTOFF of the last *Apollo* moon rocket from PAD 39 signaled the "end of the beginning"† of the movement of civilization into space. With the fire, tears, and beauty of that early-morning launch from the John F. Kennedy Space Center, humankind stood on the brink of an indefinite period of exploration, settlement, and progress beside which even the history of the

* NASA transcript of the *Apollo 17* mission. † Gene Cernan's motto for the *Apollo 17* mission.

Americas may ultimately pale. The Apollo adventures had become humankind's first halting but clearly personal look at its universe.

Carried forward in many ways, this adventurous character of humans uniquely identifies us among the known species of nature. We have the audacity to try to understand our place in the universe and in its future. We have the further audacity to try to understand, preserve, and benefit from the Earth, now with a corrected vision from space. Ultimately, we expect, as a species, to use this understanding to alter the universe and to better our place within it.

But then, as we leave the familiar, there comes a strange new perspective of the Earth filling only one small window and gradually not even doing that. No longer just the comfortable planet of our past, the Earth appears as only a fragile blue globe in space.

Like our childhood home, changing in time but unchanged in the mind, the jewel-like Earth revolves beneath us in the infinite blackness of its setting. No apparent purpose governs the cyclic wandering of Spaceship Earth through its own universal sea. The forces that hold us to this invisible course have been long explained but never understood. They produce waves we have not felt, storms we have not seen, and history we have not read. Only the certainty of continuity has been a reality during our voyage around the sun, through the galaxy, and across the universe.

For 3 days, the fascinating changing scenes of an ever smaller Earth penetrated our thoughts, until outside, a dark looming presence increasingly made itself felt as much as seen. The disk of the black moon grew in aspect, blocking more and more of the star field. The grazing sun illuminated only a thin eastern arc, and soon even that disappeared as we flew out of sunlight into lunar shadow. Finally, as the spacecraft curved around the moon, the mountains of the lunar highlands silhouetted briefly against distant cloud patterns before eclipsing the entire Earth. With this moment, the first loss of communications with Mission Control had begun, and the inevitable quiet of the lunar far side enveloped us.

A few minutes later, *America*'s rocket engine burned flawlessly for almost 7 minutes, and we were in lunar orbit. We looked first toward the airless sunrise and saw increasingly bright and glowing streamers of the corona radiating vast distances from a hidden sun. A faint, thin, glowing halo, possibly from dust just above the lunar horizon, joined each streamer and faded gradually to the north and south of our orbit. Then, just before sunrise, the center of each streamer became a blade of brilliant light projecting above the dark lunar horizon like a luminous sword. A few high peaks cast their reflected light from the horizon's edge across the dark highlands and then—sunrise!

At last, held to our own cyclic path around the moon, we saw the Earthrise, a lasting emotional and artistic reward for a generation's spirit, daring, and imagination. The horns of the crescent sapphirine Earth contrast with the edge

From space, we have a strange new perspective of the Earth appearing as only a fragile blue globe.
NASA

of the bright lunar horizon as a jewel contrasts with a candle's flame. A friend from home spoke to us again, breaking the spell of the silent far side with the same divided pleasure that comes as a whisper breaks the delicious unconsciousness of morning.

Having seen the Earthrise over the black-framed lunar horizon, we reaffirmed humankind's evolution into the universe, never again to be satisfied with only the beauties and comforts of the rising Earth. Paradoxically, we also found enhanced awareness and appreciation of all we are now willing to leave behind.

That lonesome marbled piece of blue with ancient seas and continental rafts will be our ancestral home as new families settle the solar system. The modern challenge, emphasized by this new view of Earth, remains to both use and protect that home while our children spread outward among the cosmic islands of space. If we are successful, and if freedom survives as well, the historians of the solar system may decree that for this we shall be most remembered.

A SPECIAL PLACE

A*POLLO 17* visited another special place in the solar system, a valley on the moon now known as the Valley of Taurus-Littrow—a name not chosen with poetry in mind. However, as with many names, events create the mind's poetry—events surrounding not only 3 days in the lives of three men,

The first view of Earthrise rewarded a generation's spirit, daring, and imagination.

NASA

but also the close of an unparalleled decade in human history. During that brief span of time, human beings left the Earth to live on and explore another world.

The valley has been largely unchanged by our visit, while change has governed the civilization that sent us. The valley has been less altered by being explored than have been the explorers, less affected by all we have done than have been the millions who, for a moment, were aware of its towering mountain walls, its presumptive human visitors, and then its silence.

Viewed initially from the unmarked mathematical paths in orbit 10 miles above, Taurus-Littrow's cratered face took on a changing personality: first the dark, unrelenting cold of a lunar night; then the forbidding, starkly defined streaks of sunrise shadow and light; and, finally, the softer morning contrasts that nonetheless foretold the impending harsh, desertlike glare and heat of lunar noon. The sunlight's variable mood on the still distant valley floor gave multiple aspects to the meteor-torn craters Camelot and Cochise, the avalanche-blanketed Jefferson-Lincoln Ridge, and the dusty Tortilla Flats.

But now the time had come to see for ourselves. Training, procedures, and the rush of planned events masked the potential finality of separation of *Challenger* from *America* before descent into the valley. Oblivious to such thoughts and leaving Ron alone in *America*, Gene and I completed preparations for igniting *Challenger*'s descent rocket engine for our westward flight down

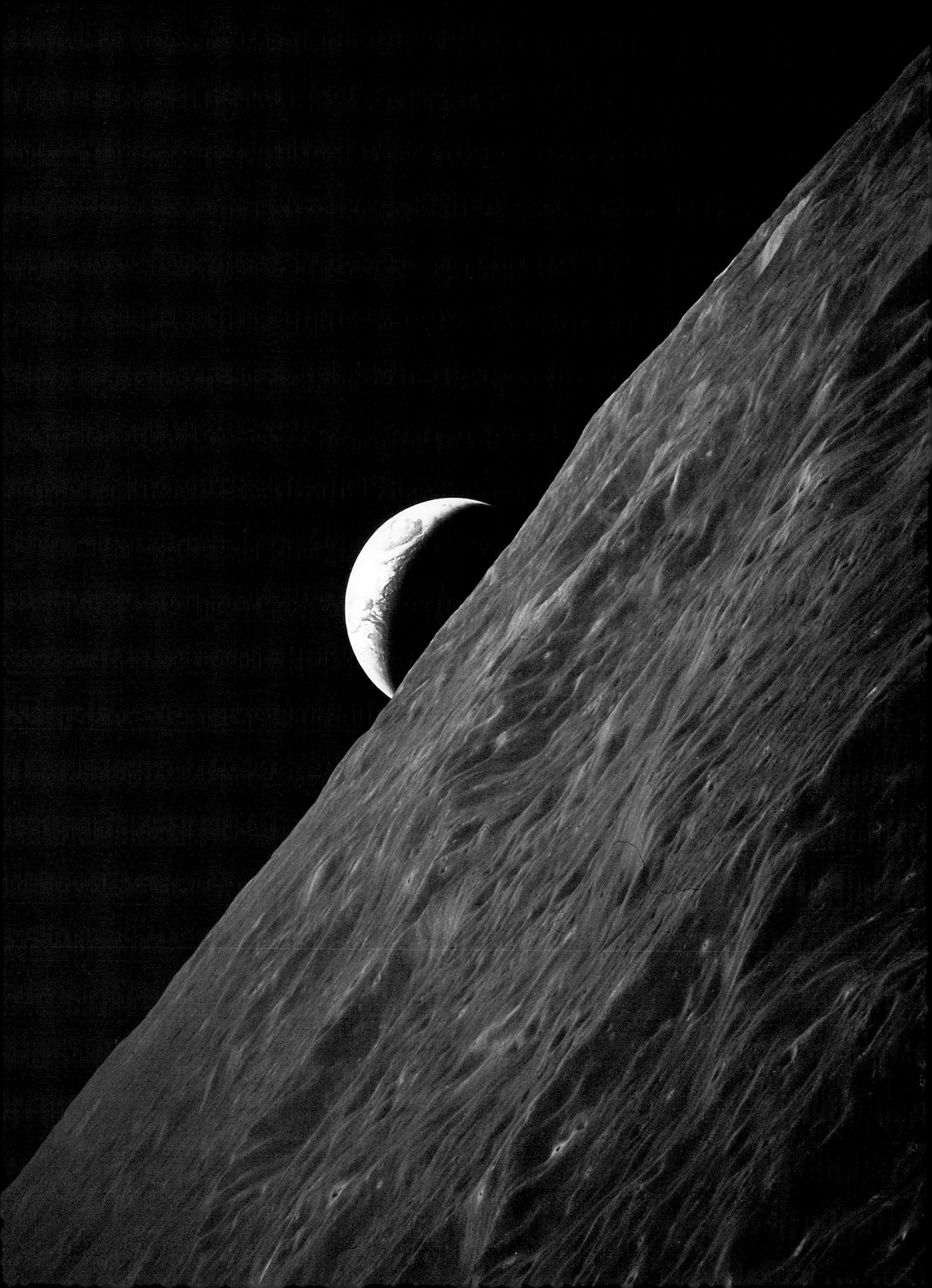

into the valley. With communications reestablished at Earthrise and the final refined navigational data loaded into the guidance computers, the descent rocket ignited. Aligned along our orbital path, the rocket's thrust fought against the speed of the spacecraft. In effect, we began to fall out of lunar orbit.

Several miles from touchdown and about a mile above the valley floor, *Challenger* pitched forward right on schedule. Now Gene could use numbered crosshairs mounted on his window to monitor the computer's course toward the planned landing point. "And there it is, Houston. There's Camelot! Wow! Right on target!"* My own hurried glance showed that friends on Earth and Ron orbiting above had not failed as distant navigators. The triangular windows revealed both the familiar pattern of craters ahead and the bright, sloping mountain ramparts to the sides of our new home. Then, the press of activities required for landing took charge again.

* NASA transcript of the *Apollo 17* mission.

The Valley of Taurus-Littrow was less affected by exploration than were the explorers and the millions who, for a moment, were aware of its towering mountain walls, its presumptive human visitors, and its silence.
NASA

At about 500 feet above the surface, using velocity and altitude figures that I read aloud from the computer display, Gene slowed *Challenger*'s descent manually and began to maneuver away from boulders and craters that could interfere with a stable touchdown. As we went through 100 feet, streaks of dust, rushing away from the soundless power of the blazing rocket, radiated across the surface beneath us and out of sight. After one anxious moment when our rate of descent briefly went to 15 feet per second, a blue light flashed on the instrument panel when long contact probes touched the surface. "Contact!" "Engine Stop-Push." With about 2 minutes' worth of rocket propellant remaining, the moonship dropped with a thump into the residual streaming dust and on to the valley floor. "Okay, Houston. The *Challenger* has landed."* Slightly less than a second later, news of our landing arrived at Earth.

The elation of a successful landing and the anticipation of what was to come spread through *Challenger*'s cabin by look and action. However, my first words from the moon were, "Okay, Parker valves," as the discipline of training continued to hold. By these words, I told Gene and Mission Control that I had started the planned, open–close cycling of valves that feed fuel and oxidizer to our 16 attitude-control rockets. As we proceeded through the post-

* NASA transcript of the *Apollo 17* mission.

landing checklist, waiting for word that we would stay through at least one rendezvous opportunity (2 hours), excitement over what we had accomplished increased rapidly. I said, "Batteries look good. Oh man! Look at that rock out there!" Gene: "Epic moment of my life!" I: "Where'd you land? You never let me look outside at all. Hey, you can see the boulder tracks!"*

My first view out the right-hand window, looking northwest across the valley at mountains 2000 meters high, encompassed only part of a truly breathtaking vista and geologist's paradise. Only later, when I could walk a few tens of meters away from *Challenger*, did the full and still unexpected impact of the awe-inspiring setting hit me: a brilliant sun, brighter than any desert sun; fully illuminated valley walls outlined against a blacker-than-black sky; with our beautiful, blue and white marbled Earth hanging over the southwestern mountains. And we were there and part of it.

The relentless clock moved on. We rushed to safe the spacecraft systems and, while describing the view ahead, impatiently awaited the words that would direct us to proceed with the planned mission or, heaven forbid, send us immediately back into orbit because of some technical problem. The relative security of *Challenger* now seemed as confining as a classroom on the last day of school. This now familiar, yet still untouched, new world beckoned.

* NASA transcript of the *Apollo 17* mission.

Slightly over 11 minutes after landing, although it seemed much longer, Mission Control said we were "stay" for our first 2 hours—that is, until *America* was again in position for a direct rendezvous opportunity. A little later, approval was received to prepare for the planned extravehicular activity (EVA). Our tasks included reconfiguring *Challenger*'s systems, realigning the guidance platform with star sightings, describing the lunar features visible from the windows, eating our first meal since breakfast with Ron, and preparing our spacesuits for our first excursion outside. Four hours after landing, we had checked the spacesuits, placed a water-bag nipple and a fruit stick within reach of our mouths, attached a urine-containment bladder, and filled a bag with cameras and other gear while impatiently awaiting the word to depressurize the cabin.

"You'll be glad to know you are GO for depress," CAPCOM Bob Parker finally reported.* With the cumbersome spacesuits zipped, latched, and checked and the spacecraft open to the airless western reach of Taurus-Littrow, we worked our way through the yawning hatch of *Challenger*, lumbered down its ladder, and touched the gray dust of eons. Gene touched first and dedicated "the first step of *Apollo 17* to all those who made it possible." I added, "[T]he next generation ought to accept this as [a] challenge. Let's see them leave footprints like these some day."†

* NASA transcript of the *Apollo 17* mission. † NASA transcript of the *Apollo 17* mission.

With the first steps and words in the nearly 4-billion-year-old valley complete, preparations for 3 days of exploration obscured the newness while we worked with familiar gear near the spacecraft. The nature of our activities, the restriction on our peripheral vision caused by the helmet of our spacesuit, and familiarity born of so many months of testing and training initially limited the emotional and visual impact of the valley. Then, the schedule of planned tasks let me move away until the brightly reflective, spiderlike *Challenger* became only a part of the total magnificent scene.

One of the most majestic panoramas ever viewed by humankind confines the Valley of Taurus-Littrow. Toward the open, western mouth of the valley, beyond the surprisingly close curve of the horizon, lies the center of the vast basin plain of Mare Serenitatis. Close by to the north and south, the roll of gray hills across the valley floor blends with bright slopes that sweep evenly upward, tracked like snow, to the rocky tops of the massifs 2300 meters above. Behind and to the east, the knobby irregularity of a highland wilderness climbs into a low, blinding sun.

The Valley of Taurus-Littrow does not have the jagged youthful majesty of the snow-draped, thrusting Himalayas; the layered, water-carved canyons of the Colorado; the symmetrical, glacially hewn fjords of the north countries; or even the mysterious, intriguing rifts of Mars. Rather, it has the subdued and

A brilliant sun, brighter than any desert sun, illuminated valley walls outlined against a blacker-than-black sky. And we were there and part of it.

Overleaf: NASA

ancient majesty of a chasm whose cataclysmic origins have been obscured and softened by time.

The massif walls of the valley rise to heights that compete well among other natural wonders of the planets, but they rise and stand with an unconcern that belies dimensions and speaks silently of 4 billion years of continuity in the scheme of evolution. Still, the valley's silence remains transitory. Cliffs yet release massive boulders to roll and bounce down dusty slopes. Heat from below and from the sun warms the valley floor as rare meteors spread new chapters of creation in glass and crystal. Soils store energy resources adsorbed from the solar wind. Craters continue as the archives of the life of the sun.

Taurus-Littrow looks like a valley in high and rugged mountain terrain, above tree line, subdued by a deep fall of new snow. But instead of new snow, meteor-pulverized rock, mineral, and glass debris covers the waterless valley and plantless mountains. A brilliant sun, as bright as any Saharan sun you can imagine, illuminates the gray blanket of sparkling dust. However, black shadows and a black deep-space sky make a contrast in scene that startles eyes accustomed to Earth's gentler views.

The first EVA, just over 7 hours long, saw the science station deployed and activated, as well as our first geologic studies and sampling of craters in the titanium-rich basalt lying beneath the valley floor. The science station includ-

ed experiments to investigate gravity-wave-induced oscillations of the whole moon; seismic activity caused by moonquakes, meteor impacts, and our own explosives; the composition of the thin lunar atmosphere; micrometeorite impact rates; heat flow from the lunar interior; and neutron and cosmic-ray fluxes.

On the second EVA, of 7 hours and 30 minutes, the battery-powered lunar rover took us 7 kilometers away from *Challenger* to the South Massif at a speed of about 10 kilometers per hour. This may not seem very fast, but when you hit a bump in one-sixth gravity at this speed, you spend the next 10 meters off the ground. We sampled boulders at the base of the south wall of the valley and explored a broad area of rock debris from an avalanche that had come from the steep side of the massif. Our surprise discovery of orange soil in the rim of the 80-meter-diameter impact crater Shorty also occurred during this period. This chemically unusual material, from 3.5-billion-year-old volcanic fire fountains, has given new insights into the origin of the moon and the nature of rocks 500 kilometers below its surface.

The third EVA, which lasted 7 hours and 15 minutes, provided an opportunity to study large boulders that had rolled and bounced down the north wall of the valley. From them, we learned about what happens when large objects from space hit, break, and partially melt planetary crusts. On Earth, impacts of this magnitude appear to cause large-scale extinctions of life forms. During the detailed examination of one very large boulder, the unexpected discovery of a subtle contact between two types of impact-generated debris, one intrusive into the other, again proved the worth of the trained human eye in exploration.

Questions often arise as to whether robotic exploration of the moon or any other planet would be less expensive than human exploration and provide all the essential scientific return. This question, of course, can never be answered to everyone's satisfaction, if only because of sincere disagreements over what constitutes "essential science." Clearly, robotic systems will and must make increasingly important contributions to space exploration; however, the spontaneous human observation, integration, and interpretation of the total dynamic situation in exploration activities, and a calculated human response to that situation, will be as irreplaceable in the future as throughout the past.

In our explorations of the valley, we drove the rover about 35 kilometers, collected and documented over 110 kilograms of moon rocks and soils, and took more than 2400 photographs. Gene used a rock hammer (the grip fit him better than me) and a set of tongs for most of his sampling, while I preferred a long-handled scoop with an adjustable head angle. We both had chest-mounted 60-mm electric-drive Hasselblad cameras, his loaded with color film and mine with black and white for subsequent photometric measurements of materials in place.

During detailed examination of a very large boulder, an unexpected discovery again proved the worth of the trained human eye in exploration.

NASA

At most sampling sites, we generally worked together, using a specific sampling and documentation routine that was more efficient than working alone. While Gene dusted off the equipment on the rover, I looked over the sampling area, giving a general description of the geology and what we would try to do. Then I took a down-sun photo while Gene took a cross-sun stereo-pair of photos of the area to be sampled. One of us collected rock or soil samples while the other held open a numbered Teflon sample bag to receive the samples. After stowing the bagged samples in larger bags mounted on our backpacks, one of us took both a post-sampling photo to show which samples had been collected and a circular panorama that included the site. A running commentary on each step in the operation as well as on other observations about the geology of the valley accompanied this process. The rover color-television camera, operated remotely from Earth, followed most of our activities, providing both additional scientific documentation and many humorous clips of our pratfalls to delight future audiences.

One-sixth of Earth's gravity made exploring the moon relatively easy, even in the confines of the bulky spacesuit. Although my own weight plus the total weight of the suit and life-support system totaled about 370 Earth pounds, the weight on the moon was only about 61 pounds. While exploring the Valley of Taurus-Littrow, we walked and then ran across what seemed like an infinite trampoline, where one-sixth gravity made falling feel like

being a kid again. If we did start to fall, we just rotated catlike in space, put out our hands, and, on impact, gently pushed back to a standing position. However, if the little beads of glass in the lunar soil got between a boot and a flat rock, we churned and spun with all the grace, but none of the control, of a ballet dancer. If you wonder why the astronauts seemed to enjoy working on the moon, it was because it was much easier than practicing in the cumbersome spacesuits on Earth, particularly as we had water-cooled underwear to remove body heat.

We could run with the motion and ease of a cross-country skier, gliding just above the surface for many yards at a stride, and at a sustainable rate of about 7 kilometers per hour. With no air resistance and no sliding contact

with the surface, each coordinated toe push produced an ever longer reach. Stopping exactly where we wanted to became the only concern.

Gene and I spent about 75 hours on the moon, 22 of which involved extravehicular activity outside *Challenger*. The vehicle provided one of the more serviceable and comfortable camps in my experience as a field geologist. The lower half, or descent stage, contained not only the big rocket engine that set us down in the valley, but also the equipment we'd use during our exploration. In addition, the descent stage held the food, water, and oxygen needed for the 3-day stay. The upper half of *Challenger*, or ascent stage, not only would take us back into orbit, but also provided a bedroom, kitchen, dining room, and bathroom.

Although two large, empty spacesuits made the cabin cramped, sleeping in one-sixth gravity provided better rest than on Earth—just enough gravity to

Spontaneous human observation, interpretation, and calculated response are essential activities in exploration.
NASA

feel the hammock beneath you, but not enough pressure to cause you to toss and turn. The freeze-dried, dehydrated, and irradiated foods tasted fine, certainly better than some food prepared by assistants in geologic field camps I have known. Possibly most important, no flies or mosquitos bothered us.

With our explorations over and a good rest behind us, the focus shifted to going back into orbit for a rendezvous with Ron and *America*. Preparations included review of seemingly endless contingency procedures to be used in case the normal computer-driven liftoff, ascent, and rendezvous sequences failed. So many engineering precautions had been designed into *Challenger*, however, that there was little likelihood we would be stranded on the moon. Indeed, I do not believe that any Apollo crew seriously contemplated what they would do in that eventuality. For example, among the hundreds of parts making up the ascent engine, only the exit nozzle and the massive fuel and oxidizer injector ring had no matching pair or backup component (no one could figure out a way these parts could fail!). As a last resort, we could even wire the descent batteries to the circuit breakers controlling the ascent engine fuel and oxidizer valves and force the valves open. As these pressurized liquids react on contact when they mix in the engine, we would be on our way.

We barely noted the .5-g acceleration of liftoff and the slight oscillation during ascent, partly because at the instant of ignition, the uplink communications turned into raw static. Later, we learned that a mix-up on a transfer between Earth transmitting stations had caused the problem. As we flew back into orbit on a direct rendezvous path toward *America*, I spent the first few minutes trying to restore *Challenger*'s communications while Gene monitored the guidance systems and yelled at me to "get the comm back!" (It turns out that nothing I could do would have helped. Mission Control finally restored communications.)

Its first visitors left Taurus-Littrow with an underlying sadness, even though the friends and flowered hills of Earth awaited our return. Maybe we thought of discoveries left unmade and tasks left undone; maybe we considered the more than 7 years of commitment now over. Nonetheless, during the liftoff from the valley and the subsequent rendezvous with Ron over the still unexplored lunar far side, regret mixed with my feelings of an adventure well done. Tears of farewell came later as our friend *Challenger* left without us for its second and final journey to the valley and a controlled crash for seismology near the South Massif.

RETROSPECTIVE

THE ONLY strong sense of unreality in our work on the moon came from the image of home, the Earth, always hanging in the same spot in that black sky, 230,000 miles away. This marbled blue and white globe, with its red and orange desert beacons, will remain the most beautiful of homes we will have in our solar system. For those who venture to Mars and beyond, even the unreality of the Earth above them will disappear as it shrinks to a point of bluish light near a setting or rising sun.

Walking on the moon *really* feels like one of those experiences in life that remains forever meaningful for one specific reason: although the experience has been anticipated through pictures, books, the accounts of others, even computer simulations, when the actual event stimulates our feelings, nothing can prepare us for the emotions and perspectives of actual personal experience.

Standing on the rim of the Grand Canyon for the first time, reaching the summit of Mount Everest, or experiencing the first true love, the first religious awakening, or the birth of a child stand as such experiences for many. Through them, life itself becomes increasingly meaningful. The essential human ingredient in such experiences is, of course, "being there."

"Being there" in the middle of the vast, majestic Valley of Taurus-Littrow, being at the foot of the towering lunar Appennines, or being on the historic plains around Tranquillity Base have taken their places among the uniquely meaningful experiences of humankind. Only 12 men, however, have visited these places, and each has experienced them in his own special way. Although each astronaut has shared his trip to the moon with millions of television viewers on Earth, he has individual memories of the space adventure as well as a uniquely personal view of Earth and beyond.

"Being there" adds the human element to life's events. The desire to "be there" will continue to drive young people away from the established paths of history on Earth and to the planets and the stars. Yes, they probably will follow the examples of previous explorers and offer "practical" or "pragmatic" reasons to rationalize going "up into space in ships"—a route to the Indies, first to reach the poles, beat them to the moon, or get the helium before they do—but it still will be a rationalization for the basic human desire to "be there."

This transcendent desire has carried humankind from the caves to the moon. "Video pictures and data streams" from Mars, no matter how good or how complete, will never be enough for the parents of the first Martians. For that generation of pioneers, some of whom live among us today, the moon will become merely a way station to greater adventures, as were the ports of Spain for Columbus and the Spanish explorers who followed him and Franklin, Missouri, for the explorers of the American West.

The Valley of Taurus-Littrow has been part of the unfolding of thousands of millions of years of time. Now, it has dimly and impermanently noted humankind's homage and footprints. The return of humans is not the concern of the valley—only the concern of humans.

EDWARD C. STONE

Voyager: A Journey of Exploration and Discovery

THERE are striking similarities—and equally striking differences—between the voyages of exploration in the sixteenth and seventeenth centuries and the programs of planetary exploration of the past 30 years.

Columbus, Magellan, Vespucci, Drake, and others set sail from Europe 500 years ago to destinations about which they knew little. But they were determined to explore, and so they went.

Over the past three decades, the National Aeronautics and Space Administration (NASA) has launched a series of robotic explorers—*Mariner, Viking, Pioneer, Voyager, Galileo, Magellan,* and *Mars Observer*—to seven of the eight other planets of the solar system. Unlike the earlier voyages of exploration, we began ours with the knowledge gained from several centuries of observations of the planets with ground-based telescopes. Even so, the first close-up images and measurements of the surfaces, atmospheres, rings, moons, and magnetospheres proved to be just as surprising to us as the flora, fauna, and inhabitants of the New World were to the European captains and crews hundreds of years ago.

The Voyager missions, which originated at the Jet Propulsion Laboratory (JPL) in the mid-1960s, became an approved NASA project in 1972. Those of us fortunate to be on the Voyager team knew that we were embarked on an exceptional voyage of exploration when the twin spacecraft roared aloft from Cape Canaveral in 1977, but none of us expected the wealth of discov-

Photomosaic of the four outer planets of the solar system, approximately to scale, as imaged by the Voyager *spacecraft between 1979 and 1989.*
Top to bottom: Jupiter, Saturn, Uranus, and Neptune (NASA)

ery that was revealed by the 12-year journey to the outer reaches of the solar system.

The same year the project was approved, I accepted the role of project scientist. As the chief scientist, I integrated the efforts, goals, and plans of more than 100 scientists organized into 11 investigation teams—Imaging Science, Infrared Radiometry and Spectrometry, Photopolarimetry, Ultraviolet Spectrometry, Radio Science, Magnetic Fields, Plasma Science, Low-Energy Charged Particles, Cosmic Rays, Plasma Waves, and Planetary Radio Astronomy—selected competitively by NASA. Each team was led by a principal investigator or team leader, and these individuals constituted the Voyager Science Steering Group under my chairmanship.

It may not be readily apparent, but a planetary encounter is actually a set of highly integrated scientific investigations executed under tight time limits. Unlike an experiment in a campus laboratory that might unfold over months as professors and graduate students set up their apparatus, study their initial results, and then modify their approach to get better answers or pursue unexpected developments, a planetary encounter is an inquiry compressed into a handful of days.

With a probe closing in on a planet at a speed of more than 30,000 miles per hour, mission scientists have just one opportunity to make the right observations. Everything must be planned in advance, and there are no second

Teams of people working in the control center, supported by others in research labs and aerospace firms, stay in contact with robotic explorers, sending operational commands, handling problems, and receiving data. This intense behind-the-scenes human effort is critical for mission success.

NASA

chances as the spacecraft hurtles toward the next planet. Also, unlike in the laboratory, the discoveries and puzzling observations during an encounter occur in such rapid succession that there is little time to contemplate their significance before attention is drawn to the next new result.

I was responsible for coordinating the initial reports of the observations and discoveries of *Voyager 1* and *Voyager 2*. During each planetary encounter, the team of scientists assembled at JPL reviewed the latest observations so that we could immediately share the most important and exciting results with the public through television and print, even though it might take us months or years to fully comprehend their import.

As the chief scientist, I also mediated between the desires of the scientists and the assessment by the engineers of what was feasible. This often involved a judgment as to the priority of the proposed observations, the degree of observational complexity that was achievable, and the level of risk that was acceptable.

In planning a mission, scientists naturally urge that the spacecraft and observational sequence be designed to make as many measurements as possible. This impulse is understandable, considering that the mission may be the only chance an individual has to explore a particular planet.

The engineers who build and operate the spacecraft, however, must consider such constraints and risks as mass and power restrictions; component life-

times; navigational uncertainties; data gathering, encoding, and transmission; the availability of ground-based receiving antennas; computer capacities; the relative positions of Earth and the target planet 5 or 10 years hence; the planet's radiation environment; the number and location of its moons; and, of course, funding and schedules. Although the engineers would like to accommodate all scientists' requests, that is not possible and so trade-offs must be made.

Fortunately, the Voyager missions' science and engineering communities worked together well. We managed to develop a set of scientific objectives—initially for Jupiter and Saturn, and expanded later for Uranus and Neptune—that

In planning a planetary encounter, a team of investigators works together to make decisions about research priorities. As Voyager 1 *approached Saturn, for example, the team considered alternative sequences for observing the moons, the rings, and the planet itself.*

NASA

were both comprehensive and definitive, ambitious yet feasible. Let me cite a specific example:

- The trajectory of *Voyager 2* was deliberately chosen to be farther from Jupiter than that of *Voyager 1*, in case the radiation environment was more hazardous than anticipated. But we had to make a choice. Should we fly *Voyager 2* close to the Jovian moon Europa, the only one of the four large moons not observed close up by *Voyager 1*, and obtain high-resolution images of its bright icy surface? Or should we fly the spacecraft behind (as seen from Earth) another moon, Ganymede, so its ultraviolet spectrometer might determine if the moon had, as some had suggested, a thin atmosphere? We could not do both.

In the Science Steering Group, we decided to forgo the search for Ganymede's atmosphere so that we could complete the survey of all four Galilean moons (the others are Io and Callisto). Our choice was confirmed by subsequent ground-based observations that found no traces of an atmosphere on Ganymede, while the *Voyager 2* images revealed that Europa's icy surface is the smoothest in the solar system, suggesting the possibility that the ice is floating on an ocean.

We had to make a similar trade-off at Saturn:

- As *Voyager 1* began its long approach to Saturn, we had to choose the tim-

Minimizing risk is an important factor in planning space missions. Uncertain about the extent of Saturn's rings, Voyager 1 *mission planners working out the trajectory had to weigh the possibility that the spacecraft could be damaged by dust particles beyond the visible rings. Images showing the complexity of these rings were among* Voyager's *most notable results.*

NASA

ing of our encounter with its moon Titan. Should it be before or after our closest approach to the planet itself? Titan, almost as large as Mercury, was long known to have a substantial atmosphere, perhaps as dense as Earth's and possibly resembling in some ways our own primordial atmosphere 4.5 billion years ago. The reason for flying by Titan before the closest approach to Saturn sprang from uncertainties about the extent of Saturn's rings; if dust and debris spread beyond the visible rings, then there was some chance that the spacecraft could be damaged by particle collisions as it flew by Saturn, possibly leaving it inoperative for any subsequent encounter with Titan. Flying by Titan before Saturn was safer, but that trajectory precluded the optimum geometry for probing the thickness of Saturn's rings with the spacecraft radio beam, another priority objective. Recognizing the advantages of a Titan-before-Saturn encounter, I explored with Len Tyler of Stanford University, the team leader for Radio Science, the details of the radio observations. Fortunately, he found that some Titan-before-Saturn trajectories would provide the right geometry for study of the rings, so we could accomplish both the Titan and the rings objectives with minimal risk.

Although we tend to talk most about scientific success of the Voyager missions, it was also a remarkable engineering accomplishment. When I joined the project in 1972, I had little appreciation for the importance of systems

The Voyager *spacecraft and other robotic explorers have starred in an era of unprecedented discovery.*

NASA

design. It soon became clear, however, that under H. M. Schurmeier's leadership the project team was carefully considering the capabilities of each subsystem so that the total system would perform as required with reliability, robustness, and flexibility. Although the two spacecraft were designed for 4-year missions to Saturn, the system design allowed us to extend the reach of *Voyager 2* out to Neptune, three times farther away from Earth.

Two examples illustrate the system's robust design:

- *Voyager 2*'s scan platform rotates about two axes to aim the camera and three other science instruments. Unfortunately, the platform jammed shortly after the spacecraft made its closest approach to Saturn in 1981. The cause was thought to be a loss of lubricant in a gearbox bearing caused by extended periods of high-speed motion. This was of great concern because *Voyager 2* was on its way to the first flyby of Uranus without a way to point its camera and other instruments at the planet. Although no one had anticipated this specific problem, the system was sufficiently robust to provide us with several options: changing the drive power to the platform's drive actuator, varying the temperature of the balky gear box, reducing the rate at which the platform moved, or leaving the platform parked and turning the spacecraft to make observations. We used all these options to free the jammed actuator and to achieve our scientific objectives at Uranus and Neptune.

◆ The *Voyager* spacecraft were designed to transmit images and other data from as far away as Saturn. At Neptune, which is three times farther away, the radio signal is only one-ninth as strong. To compensate, JPL engineers took several steps. They increased the diameters of the radio antennas of the Deep Space Network's three tracking stations, they developed ways to combine the signals from several antennas to increase their strength, and they reprogrammed *Voyager 2*'s computer to compress the total amount of data being sent back to Earth.

Although the two *Voyager* spacecraft may have looked the same as when they were launched in 1977, both they and the ground-based receiving system had much improved capabilities by the end of the 1980s. Without these improvements, the wealth of images returned from Neptune would not have been possible.

Both spacecraft continue to function quite well. *Voyager 1*, which encountered Jupiter and Saturn before heading upward and out of the plane of the planets, looked back at the solar system in 1990 to take a mosaic of images of the sun and six of the nine planets. *Voyager 2*, which visited Jupiter, Saturn, Uranus, and Neptune, is now headed downward below the plane. Both are searching for the outer perimeter of the heliosphere and the beginning of interstellar space and are expected to continue operating well into the twenty-first century. Their radioisotope thermoelectric generators have enough power left to support operation until at least 2015, when *Voyager 1* will be 12 billion miles from the sun.

The discoveries made by the *Voyager* spacecraft are extraordinary: the roiling storm systems of Jupiter, the complexity of Saturn's rings, the volcanoes on Io, the variety of planetary moons and the histories recorded in the craters and canyons that mark their icy surfaces, the bizarre misalignment of the magnetic axes of Uranus and Neptune, the Great Dark Spot on Neptune, and the nitrogen geysers on the moon Triton make up just a small part of the Voyager missions' legacy.

The Voyager missions have been the journey of a lifetime for all of us who were fortunate to have been involved. Much as the explorers of centuries ago brought back to their European contemporaries word of a New World, *Voyager 1* and *Voyager 2* have revealed to everyone the many worlds of unexpected diversity that share the solar system with Earth, worlds that future generations will have the opportunity to explore in their own journeys of discovery.

CHALLENGE: VIEWS ON EXPLORATION TODAY

A PERSUASIVE argument for human exploration in space is that people learn from experience. They cope with the unforeseen, change plans, correct mistakes, and respond to opportunities with speed and resilience. In view of these perceived virtues, it is fair to inquire what lessons learned from past exploration might guide and inform new ventures in space.

The next four essays consider science, discovery, and challenges for space explorers. The writers examine attitudes, research practices, priorities, and possibilities in space. Their optimism about the wealth of knowledge to be discovered is tempered by caution. Thomas Lovejoy and Stephen Jay Gould present reasonable approaches to a scientist's dream scenario: studying a pristine environment and searching for evidence of life. Timothy Ferris urges a change in attitude, renouncing the use of space for mundane purposes and instead recognizing there the ultimate repository of beauty and wonder. Walter

Massey highlights links among exploration, discovery, imagination, and belief.

An implicit issue is the ethics of exploration—how opportunities ranging from localized fossil-hunting to global terraforming should be pursued wisely, respectfully, and responsibly. Learning from past experience, explorers of other worlds might be guided by prudent environmentalism. If exploration led to settlement, what lessons from the social history of migrations and colonizations on Earth would be instructive?

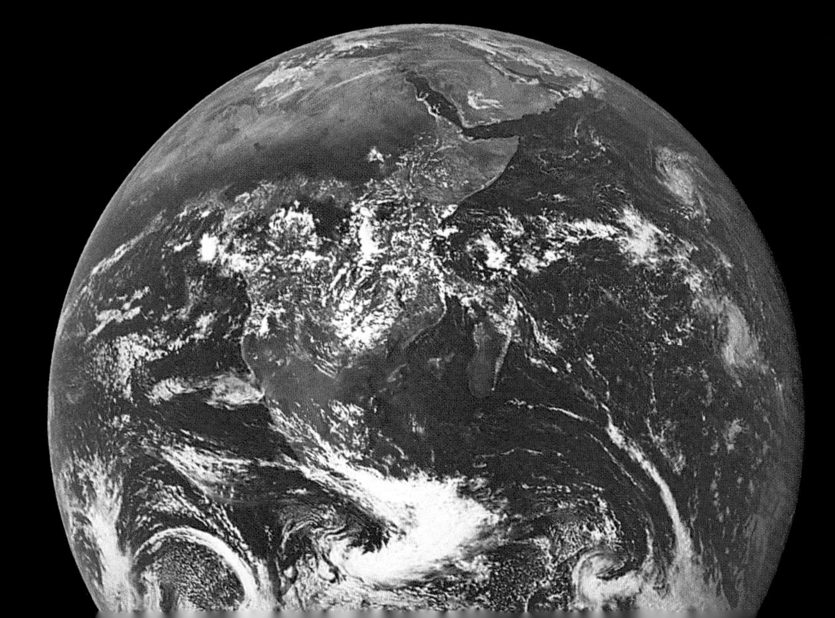

THOMAS E. LOVEJOY

Planning for a New Planet

WHEN (not if) the moment finally comes that space colonization of some sort is practical, the debate will be predictable. The argument will be predominantly environmental because ecological concerns will have come to frame all debate. As the human population swells beyond 10 billion and technology—perpetually abrim with promise—tags along, never quite able to relieve the mounting pressures on this planet's support systems and resources, ecological concerns will frame all debate.

One set of environmentalists will denounce the exercise as escapism, a waste of money, a way to ignore problems grown egregious at home, and a flight from responsibility. Another set will welcome the notion as a way to skim off a segment of the excess population, with its own potential growth, and hence relieve some of the pressure, essentially a modern argument for the frontier as a "safety valve."

Yet others will concern themselves not with the whether, but with the how. Indeed, even now there are those who dream of terraforming—making distant planets with inhospitable climates and atmospheres into habitable worlds. The "how" segment of future society will be able to ignore the late-twentieth-century debate over the uncertainty of global climate change from increased concentrations of greenhouse gases. By then, it will be obvious whether climate change is in train or whether society was somehow able to halt it.

Earth, as seen from the Galileo *spacecraft on its way to Jupiter. Space exploration has stimulated an appreciation of our planet as a unique, and vulnerable, biosphere.*
NASA

Nonetheless, it will be clear from gross comparisons of different planets and their properties (as Carl Sagan and other scientists have already noted) that adding carbon dioxide to a cold planet's atmosphere could bring it up to a reasonable average temperature for establishing colonies of organisms from Earth. Carbon-dioxide-releasing bacteria might be the first organisms introduced until there was enough carbon dioxide that green plants could follow. In the later phases of this terraforming, small colonies of people would be established based on the knowledge gained by experimental endeavors on Earth or perhaps the moon.

There will be some who question a headlong rush into terraforming. Debate will form around whether it is a feat of imagination or of arrogance, whether it is practical or impractical. Some will argue for the importance of a preliminary research phase during which the basic nature of the planet proposed for transformation can be examined. Remember, they may say, how important it was for Sir Joseph Banks to have participated in the Australian expedition of discovery. Be mindful of the importance of the explorations along the possible routes of the transcontinental railway under the aegis of Smithsonian Assistant Secretary Spencer Fullerton Baird, and of the work of the scientists Napoleon brought along to Egypt. All three examples embody the principle of coupling scientific research with exploration, of guiding human settlement scientifically.

The environment of Mars, as is evident in this view of Valles Marineris, is not hospitable for terrestrial life, yet Mars has long been a target for future exploration and possible settlement.

U.S. Geological Survey

On a lifeless planet slated for terraforming, the research phase would concentrate on understanding the physical processes that govern that world. For example, natural fluctuations in incoming solar radiation could be taken into account, and the desired atmospheric composition and consequent climatic regime, neither of which need be exactly like that which exists on Earth today, would be determined. Terraforming itself would be highly experimental. Would there be any way to predict in advance what limiting—indeed, self-limiting—factors might stop bacteria from releasing carbon dioxide before minimum desired levels are attained? Conversely, could a runaway greenhouse effect be predicted and steps taken to halt it? Once change is initiated, there is no return to the pristine state, no matter how desirable.

Once the ideal climate and atmospheric composition were achieved, the next phase would be to build a more elaborate ecosystem by adding more organisms. This would differ from the highly relevant field of ecosystem reconstruction or restoration ecology, for which the goal is to re-create a naturally occurring ecosystem so that the original species are still extant and available. For terraforming, there might be no reason not to combine species that do not naturally occur together, but to avoid unpleasant surprises it will be necessary to do a lot of contained experimental work beforehand.

As the new biosphere becomes richer in biological elements and the actual mass of living matter (biomass) becomes larger, there will come a point

when, as on Earth, its collective metabolism will trigger various fluctuations and feedbacks. For these to be even partly predictable, a great deal of careful advance experimentation on a small scale will be necessary.

If, as on Mars, there is a possibility that some forms of life have existed on the planet, it would be imperative to understand a fair amount about those life forms and the implications for a terraforming project. If the cause of their extinction could be deduced, what would it mean for the planned terraformation? Even if there turn out to be no pragmatic implica-

It may be possible to make another planet into a habitable world by changing its temperature and atmosphere, but "terraforming" raises a host of practical and ethical questions about managing global change. This artist's concept depicts a terraformed Mars with water and clouds, looking more like Earth.

Michael Carroll

tions of that immediate sort, surely it would be desirable to learn about the previous life forms. It would have overwhelming implications for biology if silica had played the role in biological construction that carbon does on Earth.

If the planet selected for colonization were abundant with life, while enormously exciting from scientific and other viewpoints, the questions about terraforming would be much more complex, especially given our experience with environmental problems here on Earth. There would be a strong argument for a thorough inventory of the variety of life forms prior to any colonization effort, although field stations of some sort would be necessary and permissible. Colonization would then be planned, with representative areas set aside for conservation of this biological diversity. These sectors would probably be designed like current UNESCO-designated biosphere reserves, which have a core of basically pristine environment surrounded by a zone available for manipulative use and study. This arrangement permits an inherently rational scientific approach: the core provides what scientists term a control, an unaltered basis for comparison with experiments in the same type of ecosystem. Such an approach would inject a significant measure of constraint into the enterprise. There is ample history on Earth of deleterious effects caused by organisms introduced into areas where they do not occur naturally. Consider but one instance: the changes in North American cities and forests resulting from the accidental arrival of Dutch elm disease. Great care, including in-depth analyses of planned colonization, would be necessary to avoid accidental introductions. Beyond this, of course, it would be necessary to understand how living and nonliving systems interact to avoid disturbing the basic cycles that maintain the variety of life on this "new" planet.

THIS IS just the kind of approach, albeit long after our appearance on Earth, that conservationists are working on today. An extremely exciting form of exploration is often given the rather dull epithet of biological inventory. What is sorely needed is an age of biological exploration in which the dimensions and variety of life on Earth are revealed. It is indeed an embarrassment that modern science has progressed so far in space exploration and molecular biology while this basic research has slipped out of fashion. Should we not be interested in the dimensions of life on our home planet, Earth? Today biological science is unable to state with certainty, within a factor of 10, how many species of living things share this planet. That is, there may be 10 times more species than we currently have recorded. Only about 1.4 million plants, animals, and microorganisms have been described; the total diversity has been estimated from a few million to 30 million or more.

A major piece of this ignorance is the tropical forests. Although they occupy but 7 percent of the dry land surface of the Earth, they are incredibly rich in biological diversity and are thought to hold half or more of all species. Some years ago, Smithsonian entomologist Terry Erwin sampled insects in the canopy of neotropical rain forests. Based on the beetle census alone, he estimated that the total number of species on Earth might be 30 million or more. Tropical forests are clearly a tremendous reservoir of biological diversity that is still mostly unexplored.

Unexplored biotic diversity can be close to home. Edward O. Wilson of Harvard University suggests that some of the organisms in a handful of soil taken from almost anywhere in the United States will belong to undescribed species, and if we include bacteria, probably most of the species will be new to science. This generality is likely to hold almost anywhere in the world. The oceans are likewise poorly explored. They are home to the greatest variety of major levels of biological organization (phyla). Just recently, Smithsonian scientists R. P. Higgins and R. M. Kristensen described a new phylum that lives among the interstices of sand grains.

The life sciences confront a significant problem in the accelerating impoverishment of our biological resources. Tropical deforestation rates pushing 100 acres a minute contribute to extinction rates estimated to be 1000 times normal. Yet this really is just the extreme example of habitat destruction that is taking a toll on biological diversity almost everywhere. Add the toll taken

The tropical forests are a tremendous reservoir of biological diversity that is still mostly unexplored. A new era of exploration here would reveal the rich variety of life on our home planet.

Smithsonian Institution, Carl C. Hansen

by the spread of toxic substances and by the distortion of biogeochemical cycles, such as those of carbon, energy, and water. These can be regional in scale, such as acid rain. They can also be global, such as the increase in greenhouse gases, which threatens unnatural climate change and further loss of biological diversity because species cannot adapt to rapid rates of change in highly modified landscapes.

One of the most serious consequences of this impoverishment of the biota is that it, in turn, implies impoverishment for the future growth of the life sciences. While there have been different life forms in the past and there will be a multitude of different forms in the future, the current variety of life on Earth is an extremely valuable sample of what is biologically possible. Biological diversity is essentially a gigantic library on the life sciences. It is rich almost beyond belief. Wilson estimates that a single chromosome of a mammal like a domestic mouse contains information, as a computer would count it, equivalent to that in all the editions of the *Encyclopaedia Britannica* combined.

Amid such extraordinary richness, one might argue there need not be serious concern about losing a few percent of extant species. Yet a library is diminished by the destruction of any of its books. The loss of the basic material on which a science is built is not a problem unique to the biological sciences; it is a problem also for social sciences such as anthropology and archaeology. That makes it no more acceptable, however. Imagine, by analogy, the reactions of astrophysicists were they to be confronted with the disappearance of whole classes of celestial bodies just when space exploration has greater potential than ever before.

The exploration of life on Earth is tremendously exciting in prospect, as it involves the continual discovery of organisms that we never dreamed biologically possible. Perhaps the most surprising example found in recent years is the biological communities clustered around the hydrothermal vents on the sea floor. They demonstrate that it is possible for entire biological communities to run on the primal energy of the Earth rather than that of the sun. They also show that it is possible for organisms to live at temperatures in excess of the boiling point of water. In stark contrast, blue-green algae live in the Antarctic ice. These kinds of organisms are the true environmental extremists that enlarge the scientific understanding of the limits within which living systems can function.

TERRAFORMING is building a biosphere de novo using elements from the one that is our and their natural home. By the time such an exercise is undertaken, it is conceivable that genetic engineering may have reached the point where it is no longer limited to working with the existing gene pool. If one or more organisms different from any created by organic evolution are needed, it may be possible to create them in the laboratory. Yet such capacity will not, I believe, ever reduce the value that the existing biota will have for terraforming, both in understanding biological systems and in providing the working elements for a new biosphere. Indeed, planning for a new planet is a compelling reason for applying the environmental lessons learned here on Earth and for practicing better biological housekeeping here at home.

S T E P H E N   J A Y   G O U L D

A Plea and a Hope for Martian Paleontology

WHEN Neil Armstrong made his first small step on the lunar surface—and happily illustrated the humanity of this technological triumph by promptly flubbing his well-rehearsed line of greeting, the first words from another world, his guarantee of space in all future editions of *Bartlett's Familiar Quotations*—I was standing in the bar of a gambling casino on the island of Curaçao. I have faults and foibles aplenty, but neither filling my gills nor emptying my pockets lies among them, and this was an odd venue for me. I was there, jammed in with half the island's population (or so it seemed), for a particular reason of necessity: the bar housed the only accessible public television for miles around. Even the casino—that ultimate 24-hour institution—stood temporarily empty, as everyone crowded around the fuzzy picture to witness the greatest pure joy of recent history.

The croupiers, the bartenders, the customers, even the manager (who initially tried to turn off the set and return the crowds to their customary activity of his profit and their loss) whooped and sighed. We became one of those temporary communities so often created by shared joy or sorrow. I particularly remember one elderly, heavy-set woman, weighted down by jewelry and makeup, jumping for pleasure and executing the most elegant little jig, as Armstrong made that giant step for all of us. We may not be a particularly intellectual species (especially the subset of *Homo sapiens* that hangs out in casinos during the wee hours); we are surely not conversant with science, either in knowledge or even by interest. But, for that hour or so, everyone put aside their immediate concerns and gave rapt attention to an omnipresent object, otherwise almost always ignored (except while spooning or crooning love's tune). For a sublime moment, we all cared passionately about the moon.

This book, as its title indicates, commemorates the most ambiguous event in the history of human expansion—so troubling because Western conquest often led to the extirpation of original inhabitants, and because the year 1492 also marked the final defeat of the Moors and the expulsion of the Jews from Spain, thereby promoting a rigidity and homogenization so contrary to the spirit of exploration and, ultimately, so destructive of great nations.

The space program—our century's obvious surrogate for Columbus (despite all the misanalogies that accompany the similarities)—is not freighted with such a depth of ambiguity. Obviously, all has not been simple joy and exultation; many of us have worried about military spin-offs, the use of ostensibly universal benefits of science for enhancing tensions of the cold war (a topic that now, and happily, has a paleontological ring), the diversion of funds that might have been used for social programs, and the substitution of hype and politics for scientific benefit in establishing priorities (manned versus unmanned, space stations versus planets).

But all this debate pales before the ambiguity of Columbus, for an obvious and particular reason: Western conquest of the Americas led to the death of peoples and the dismemberment of cultures. Weep for the dead and the enslaved, but weep also for the literal meltdown of a great artistic tradition, as exquisite works of wrought gold became ingots of mammon to fuel the expansion. This century's breakout beyond our planet encountered only lifelessness amid a geologic beauty that we have not yet seriously polluted. Nonetheless, I dedicate this essay to an argument that evidence of life (albeit from my own domain of a paleontological past) may yet be found on another planet in our solar system—and that such a discovery might furnish the greatest plum of biological knowledge in all our history. This essay is, therefore—and I might as well say so right up front—an unabashed plea for enhanced exploration and for science.

The prophet Micah spoke for practicality as well as morality when he exclaimed: "But thou, Bethlehem . . . though thou be little among the thousands of Judah, yet out of thee shall come forth unto me that is to be ruler in Israel."* The greatest successes of the space program—in science and in beauty—have involved relatively small expenses, lightweight equipment, and minimally bureaucratized lean teams of dedicated scientists. Space shuttles (and

* Micah 5:2.

thoughts of space stations) must stand in awe before the accomplishments of *Voyager*, a complex of instruments that could fit in the back of a pickup truck, but that sent crystal-clear images across light-hours of space from the most distant planet in our solar system. (Neptune, so stunningly photographed by *Voyager 2*, is temporarily the farthest planet. Pluto, usual holder of this title, circles the sun on a highly eccentric path and now lies inside the orbit of Neptune. Symbols are important, and our ability to image the most distant of known planets must rank as both a testimony and a challenge.)

For all that we have gained in knowledge since the launch of *Sputnik 1*, our increments of beauty have been just as compelling and important. I was a young teenager when *Sputnik* changed our world. I am therefore a member of the last generation to reach an age of adequate memory without knowing an image of either the back of the moon or the face of the entire Earth. I choose these two discoveries as most seminal, for they symbolize both wonder and knowledge. It must be nearly impossible for people who grew up with these images to grasp the special thrill of their first entry into our visual repertoire. What better embodiment of the cliché—so near and yet so far.

To think that people (and protopeople), from day one of our lineage, had gazed on the moon's visible face and that not one of them had ever glimpsed the other side. I was so frustrated as a child; I yearned to reach out just a bit farther and turn the damned thing around. And then, one rocket ship, one photograph, and the puzzle of all time became the common property of all people.

To think that we had developed a science of cartography to map, with stunning accuracy, the architecture of our only planetary home. We could draw the Earth's surface at any scale or detail, but no one had ever seen the entire planet for the simplest of hide-bound reasons: we could not get far enough away. The masters of photography must have shared Archimedes's mixture of joy (in the power of their technology) and frustration in his famous statement about the lever: give me a place to stand, and I will move the Earth. And the photo, when we all crowded around the newsstands to see the whole Earth for the first time, was so very beautiful. "Just like the maps," we thought, and then we laughed because we had viewed the abstraction as more real than the object itself.

Less than a generation later, as *Voyager* photographed the moons of the most distant planets, we exchanged the thrill of imaging something we knew, but had never seen, for the even more delicious pleasure of viewing objects we had seen (at least as faint specks of fuzzy light on photographic plates), but never knew. To think that a signal could be sent so far to depict the most distant major moons—Miranda of Uranus and Triton of Neptune—with a

A world now unsuited for life may yield a paleontological record of its geologic past. Space explorers could go fossil hunting.

Robert Murray

clarity that we could not exceed in a photo of Uncle George just 10 feet away! I could only think of the most famous line uttered by Miranda's prototype in Shakespeare's *Tempest*—"O brave new world." And I could only remember Wordsworth's choice of an image in lamenting that the world is too much with us, and hoping that we might recapture childhood's lost sense of wonder:

> So might I, standing on this pleasant lea,
> Have glimpses that would make me less forlorn;
> Have sight of Proteus rising from the sea;
> Or hear old Triton blow his wreathed horn.

In the happy light of these successes, and in the far colder glare of current political realities, what should (or could) be next in this modern expansion? I should say at the outset that I have no theory of inherent or predictable direction for human exploration and that I tend to regard most such proposals as romanticized hogwash. I do not think that we are impelled to expand through the universe on wings of consciousness; nor do I believe that, like the Tokugawa shogunate of seventeeth-century Japan, we are likely to undo past trends and, by conscious decision, issue a ban against all further spread. Our future is as contingent as our origin and evolutionary existence. We did not have to arise; we would never evolve again if the tree of life were replanted and allowed to grow from seed; our future, although highly constrained by what we are and

where we have been, can follow a plethora of radically distinct and quite unpredictable pathways. We may expand to inundate the solar system (if not the universe), or we may blow ourselves up tomorrow morning and never take another bus to work, much less a rocket to the moon.

But if we are to continue, and if we are to be driven by the quality of potential *intellectual* reward, then an obvious next goal does lie before us—a pearl of potential knowledge so precious that (to this biologist at least) all else must take second place. I raise again, with some trepidation but not in the old way, the perennial issue of life on Mars. I am not looking for little green men, or even for littler green plants. I want to go fossil-hunting. My viewpoint is neither original nor idiosyncratic. Many colleagues agree, and I hope that we can parlay this argument into a consensus that places the possibility of a Martian paleontology at the forefront of international efforts in space.

The issue of life on other planets has always been paramount in our thoughts about the cosmos. When we had no other tool for inquiry but speculation, views ran the gamut from Earthly uniqueness to a graded series of higher beings on heavier planets, leading right up to the celestial choirs. In his *Essay on Man*, Alexander Pope described the higher creatures of Jupiter in a striking image that depicted his greatest contemporary as an ape in their eyes:

> Superior beings when of late they saw
> A mortal man unfold all nature's law
> Admired such wisdom in an Earthly shape
> And showed a Newton as we show an ape.

For obvious and entirely legitimate parochial reasons, the issue of life on other planets remained foremost, while better understanding of geologies, climates, and atmospheres pretty much limited hope to Mars. Although shreds of doubt remain in some quarters, I shall not challenge the general consensus that Mars is presently abiotic—a conclusion strongly supported by the absence of any suitable Martian environment and, especially, by the failure of several experiments, robotically performed by our landing craft on the Martian surface, to find any chemical activity that might be attributed to living organisms.

But the present is a microsecond in geologic immensity, and the climates and surfaces of planets may change over billions of years. A world now unsuited for life may still sport a paleontological record in the archives of its geologic past; no universal Gaia ordains a permanent planetary stability once life emerges.

Martian geology offers substantial reason to suspect that life, in its simplest cellular form, may once have emerged and spread on the planet's surface. Geologists have divided Martian history, as read largely from topographic evidence, into three major phases. From oldest to youngest, they are the Noachian, Hesperian, and Amazonian. The Noachian system corresponds to the early period of intensive cratering and bombardment that characterized all bodies in the vicinity (Earth and the moon as well) and ended some 3.8 to 2.8 billion years ago. Mars was then a different world in more ways than frequent scars of impact. Water flowed extensively on the planet's surface, as indicated by abundant evidence of complex and extensive channeling on older terrains (primarily the heavily cratered uplands). Substantial oceans may have existed as well. An atmosphere then rich in water vapor and carbon dioxide may, through a greenhouse effect (now well appreciated by Earthlings for Earthly reasons), have shrouded a planet characterized by "possible warm, wet pluvial conditions."*

This early period of extensive water flow has been well recognized for

*My major source for quotations and geologic information is V. R. Baker, R. G. Strom, V. C. Gulick, J. S. Kargel, G. Komatsu, and V. S. Kale, "Ancient Oceans, Ice Sheets and the Hydrological Cycle on Mars," *Nature*, 15 August 1991, 589–594.

several years. But three objections obviously arise to an inference therefrom that life might have emerged on Mars:

1. If the history of life is so chancy, contingent, and unpredictable, why should life emerge elsewhere even if conditions were appropriate? (In this, I am potentially hoist by my own petard, since I have been a major advocate for the principle of contingency.)*
2. Even if life did emerge, why should we maintain any hope of finding fossil evidence? If proponents of life are arguing for only simple cellular structures (and I would expect to find nothing "higher" in a Martian fossil record), and not for little green men who build cities and canals, then what can a fossil record preserve and reveal?

---

* Stephen Jay Gould, *Wonderful Life: The Burgess Shale and the Nature of History* (New York: Norton, 1989).

Mars was once a different world. Networks of channels on the planet's surface indicate that water flowed there; lakes and oceans may have existed as well.

Mark S. Robinson

3. Even if conditions were once appropriate on Mars, this period of possibility ended long ago and did not persist for sufficient time to permit much hope for the evolution of life.

I do not think that any of these objections has much force and will consider them in order.

CONTINGENCY AND PREDICTABILITY

THE INTERPLAY of chance and necessity, contingency and predictability, defines the complexity and fascination of the natural world. Neither I nor any advocate of contingency views the Earth's history as entirely chancy, entirely devoid of any predictable or repeatable feature. Rather, the two cardinal phenomena of chance and necessity occupy different domains of particulars and broad patterns respectively. The basic outlines have repeatable features; the details are richly contingent. To cite the obvious analogy to human history: given our basic biology and mentality, the eventual emergence of agriculture and cities seems a fair prediction (and did occur several times independently in the Fertile Crescent, the Americas, and Asia). We might even specify, in broad terms, where agriculture might emerge and what kinds of plants might be favored. But a vast realm of contingent and unpredictable detail then unfolds. What languages will the people speak? Who will conquer whom (or will conquest be a mode at all)? What will people wear? What further technology will develop?

Martian geology offers substantial reason to suspect that life, in its simplest cellular form like these Earthly microfossils, may once have emerged and spread on the planet's surface.

For the evolution of life itself, we must answer one fundamental question: Does an origin from chemical precursors lie in the realm of predictable general pattern or in the domain of contingent detail? When I first studied this subject as a student 30 years ago, contingency was the favored view (leading to substantial doubt that proper Martian conditions could yield a high probability for life). We learned that life requires a long series of chemical steps, each improbable in itself and each necessary before the next occurs. Given such staggering improbability, life emerged on Earth only because history supplied so much time. Given enough time, to cite the common catechism, the impossible becomes probable, and the improbable virtually certain. Such an argument would debar life from Mars because the early period of appropriate conditions did not persist long enough (the Noachian time of extensive channeling ended some 2.8 to 3.8 billion years ago).

But this argument is wrong on empirical grounds. It enjoyed favor when we had no fossil record for the Earth's first 3 billion years and therefore thought that most of our planet's history was lifeless. But starting in the 1950s (and by learning to look for single-celled organisms in the right places), paleontologists began to discover an abundant fossil record during this supposedly, and officially named, Azoic (lifeless) Era. We have now found fossils of unicellular organisms in rocks dating from 3.5 to 3.6 billion years ago, the oldest known sediments that could contain evidence of life (for earlier rocks

are too altered by heat and pressure to preserve fossil remains, while no rocks on Earth are older than about 3.9 billion years, since the Earth's surface probably experienced a period of melting just before then).

What implications are we to draw from the stunning fact that life on Earth is as old as it could be? I realize that the following inference is not logically necessary, but most of my colleagues would join me (as I claim no originality for this rather obvious conjecture) in stating, as the most reasonable reading of available evidence, that if life arose so rapidly, its origin (if conditions are right) probably lies in the domain of predictable broad pattern, not in the world of unlikely contingency. In other words, given proper climates and chemistry, life probably arises more or less automatically—and quickly in geologic terms. With extensively flowing water, conditions on Mars may have been well within the range for predictable generation of life. The Noachian period of abundant water may have ended some 3 billion years ago, but this

gives life more than enough time. After all, by 3 billion years ago, life had already existed on Earth for at least 500 million years—a period fully equal to the entire history of multicellular animal life, since its inception in the Cambrian explosion some 550 million years ago.

MARTIAN FOSSIL RECORD?

IF LIFE on Mars generated only simple cells, what hope do we have for a fossil record? A clam shell makes an easy fossil, but how can a bacterial cell be preserved? Again, we can look to Earth and respond, for Mars, with that quintessential quip of our laid-back culture: no problem. First of all, the bacterial and cyanophytic cells that populate the first 2 billion years of the Earth's fossil record form larger structures that are easily and abundantly preserved. These stromatolites—mats of sediment trapped and bound by living cells—form conspicuous mounds, sheets, and cabbage heads (for so do the multilayered structures often appear in cross section), sometimes large enough to build small reefs. Second, the cells themselves can be found, usual-

If life ever existed on Mars, it might appear in fossilized stromatolites—mats of sediment trapped and bound by living cells—just as part of Earth's fossil record is preserved in these specimens.

Carolyn Russo

ly within chert (perhaps formed from gels of silica that can trap cells) and often exquisitely preserved.

Moreover, rocks of appropriate type evidently exist on Mars, for photographs have revealed stratified deposits in dry lakes and streambeds. These are the very sorts of rocks, formed in the very kinds of climates, that can (as they do on Earth) preserve the cellular remains of a variety of organisms that might have evolved on Mars.

MARTIAN ENVIRONMENT

BUT WEREN'T the appropriate Martian conditions available too long ago and for too brief a time to allow any realistic hope of life and a fossil record? Two answers to this challenge effectively remove it as an argument against Martian paleontology. First, as noted earlier, a transient period of water flow during the Noachian phase of Martian history gives more than enough opportunity. We have firm evidence that similar conditions on Earth yielded abundant life in a time undoubtedly shorter (between 3.9 billion years ago, when the first rocks formed, and 3.6 billion years ago, when the first fossils were preserved) than the transient period of appropriate conditions on Mars.

Second, the article by Baker and colleagues reports exciting new evidence that later times in Martian history—the Hesperian and Amazonian, which are usually viewed as epochs of lifeless and effectively waterless conditions—may also have featured short episodes of extensive flooding amid a general coldness and dryness that could not support life.

Baker and colleagues argue that episodes of cataclysmic volcanism occasionally produced outbursts of extensive flooding from subsurface waters that had been trapped as ice in permafrost and other sources. These floods caused widespread channeling and may have released enough water to form a transient ocean over a large part of Mars's northern hemisphere (Oceanus Borealis in their terminology). Carbon dioxide and water vapor, released by this volcanism and aqueous flow, may have unleashed a greenhouse effect and "produced widespread warm, wet conditions late in Mars history." Baker and colleagues argue that these periods of temporary flooding and oceanic formation occurred several times in Mars's more recent history. Again, by analogy to the rapid origin of life on Earth, each of these "widespread warm, wet" periods would have included ample time for an appearance of life—even if total extinction then ensued with the seepage and locking up of waters, the dispersal of carbon dioxide, and the return of cold, dry (and lifeless) conditions.

Indeed, one might conjecture (somewhat archly, but not absurdly) that Mars could be a better place than Earth to study the origin of life, for this great event happened but once on the planet we know best (or at least left liv-

Evidence of life may yet be found on another planet in our solar system. It would probably rank as the greatest biological discovery in history.

Overleaf: Pat Rawlings, *First Light* (1988)

ing representatives from only one beginning), while Mars may have experienced several cycles of origin and total extinction. Earth may be limited by the equability of Gaia.

AS A FINAL point, why should we care about some potential simple cells preserved as fossils on Mars? They cannot talk to us or tell us about the secrets of consciousness. Why are they more interesting than any one of myriad potential fascinations in space—the rings of Saturn, the extraplanetary realm of comets, the light of distant stars and galaxies? This challenge, I think, invites a stunningly simple reply, one so basic that, paradoxically, we often pass it by.

We have a legitimately parochial, primary interest in the phenomena of life; we yearn to understand our distinctness from other configurations of matter. People often ask—indeed, it is almost a catechism from those opposed to expanded research in space—why so much money should be spent on celestial pipe dreams when we remain so ignorant of so much here at home. The point is often well taken, except that the basic logic of certain problems requires knowledge from extraterrestrial sources. The distinctnesses and the basic properties of life lie firmly among these problems, for these issues cannot in principle be resolved on Earth for a simple reason rooted in evolutionary theory and the very nature of experiment.

The stunning and complex biochemical similarities shared by all living creatures on Earth indicate that all of us, from algal cell to hippopotamus, are twigs and branches on a single genealogical tree of life. We are all, in other words, products of one experiment. No matter how many terrestrial creatures we examine, from Domesday to doomsday, from bacterium to beetle, we are looking at replicates of a single origin. We can learn about ranges of possibilities within this constraint, but we cannot obtain information about most of the great questions enveloping the largest of all inquiries: What is life? How else could life be built in ways that deviate from the similarities binding us all? Are these similarities necessary ways of imbuing universal chemistry with life, or do they represent just the particular configuration (among a zillion conceivable alternatives) that Earthly origin happened to embody?

We can conjecture about these questions forever—indeed we have, in dense and learned treatises from Aristotle, to Kant, to Pogo, to Monty Python. But we cannot have meaningful information until we find another experiment, independent from life on Earth. That other experiment is as close to a holy grail for biology as anything that we could ever know or find. Call me Parsifal (and I may be a perfect fool), but this grail may lie in the sediments of Mars. Parsifal prevailed; eventually cured the wound of Amfortas, the Fisher King; and substituted light for his previously painful naïveté. Going to Mars and finding appropriate sediments is a piece of cake compared with the travails of Parsifal against the Red Knight and an associated bevy of medieval encumbrances. And may we not hope that such a mission would bring comparable light and even comparable healing in the international cooperation that such efforts require and in the humility of knowledge potentially received about a more than Earthly fellowship of life.

Such thoughts are grandiose (hopelessly prideful and foolishly pompous as well), but I am saved by my utterly selfish true motivation. I want to know before I die. This temporary configuration of neurons yearns for an answer and cannot intone its *nunc dimittis* unless an honorable effort be made in its lifetime (for I cannot demand an answer, only a try).

I would like to greet the coming millennium with our first real datum on life's generality, the only one we can actively seek in a known place (otherwise, we passively wait for something to contact us or scan a boundless immensity with SETI). Life runs in cycles, and I want to stand in that bar in Curaçao once again. This time, however, I am betting. Two bucks on bacteria, but not one penny for tribute to all the knights of convention on Parsifal's pathway.

TIMOTHY FERRIS

Raising the Roof

When you come to a fork in the road, take it! Yogi Berra

TWO DIVERGING paths lead toward the future of space exploration in America. One is venturesome and open-ended; the other, limited and more predictable. Our future as a space-faring people may depend on our choosing the right path. I fear we have already embarked on the wrong one.

The two paths reflect conflicting views of our place in the universe.

The closed view dates from the days before Copernicus, when Earth was thought to stand immobile at the center of a closeted cosmos. Space in this cosmology is a cozy cloak wrapped around our planet. In a limited universe, space flight, too, is assumed to be limited. Shooting a rocket toward the stars is like throwing a ball into the air: what goes up must come down.

The open view is post-Copernican. It recognizes that the universe is vast and that its center is everywhere. To fly into space in such a universe is to go not up, but *out*. A ball thrown hard enough need never return: it can become an Earth satellite, or an emissary to other planets or to the distant stars. In an open cosmos, the potential of space flight is limitless.

Related to these models of the universe are two contrasting outlooks on life and learning. The open outlook holds that we have a lot to learn; it eschews dogma, and hostages all beliefs about the natural world to the verdict of exploration and experimentation. The closed outlook holds that we already know most of what we need to know: absolute wisdom may be found in the Bible or some other holy book or in the decrees of an authority figure, and our main

The closed view of the universe originated centuries ago in concepts of a finite, boundaried domain. Copernicus challenged the basis of this view, and modern cosmology is a story of changing views of the structure of the universe.

Library of Congress

duty is to adhere faithfully to what we have already been given to understand.

The closed, Earth-centered outlook ruled the Western world for a thousand years. Its decline began when Copernicus, who was 19 years old when Columbus first sighted America, theorized that planetary motions could be predicted more accurately if one began with the premise that Earth orbits the sun instead of the other way round. The Copernican cosmology had great impact; the word "revolutionary" acquired its social implications from the title of Copernicus's book *On the Revolutions*. And this was true not solely because of the theory's scientific content, but also because it demonstrated that ancient dogma could successfully be challenged by new ideas.

The scientific revolution that Copernicus helped set in motion led, within the next few centuries, to the realization that the sun is but one among many stars in a vast universe, and that knowledge gleaned on Earth therefore represents but a fraction of the whole. The lesson of science ever since has been that our ignorance dwarfs our knowledge; as the physician Lewis Thomas put it more than a decade ago, "The greatest of all the accomplishments of twentieth-century science has been the discovery of human ignorance."*

This news has yet to permeate the corridors of power. Many of our political leaders still talk as though space were a kind of attic, a finite high ground to

* Lewis Thomas, "Debating the Unknowable," *Atlantic Monthly*, July 1981, 49.

be occupied for privilege and profit. "Space is a place," they say. They speak of the "conquest" of space. But space is not a place; it is *all* places. We live in space—always have—and while we may explore it, we are never going to conquer it. Neither is anybody else. There is just too much of it.

Out there lie marvels that beggar our much vaunted powers of imagination—black whirlpools that swallow light, lightning bolts larger than galaxies, atomic nuclei hefty as stars, gravitational lenses that dapple distant starlight, icy comets older than the sun. Spectacles like these instruct us in the workings of cosmic processes that gave rise to our planet and our species. In doing so, they remind us that we are subject to universal law, and we are students, not rulers, of the cosmos.

Much of what we have learned about the universe has been exhilarating. But much, too, has been unsettling and has begun to make us realize how lucky we are to be here at all. Mars was once a planet of oceans, lakes, and scudding clouds. Then something dreadful happened, and today Mars is a frozen desert. Nobody knows why. Venus, too, may once have had oceans. Today it is a fur-

Today, the closed view persists in concepts for practical uses of space as a place to be occupied.

NASA

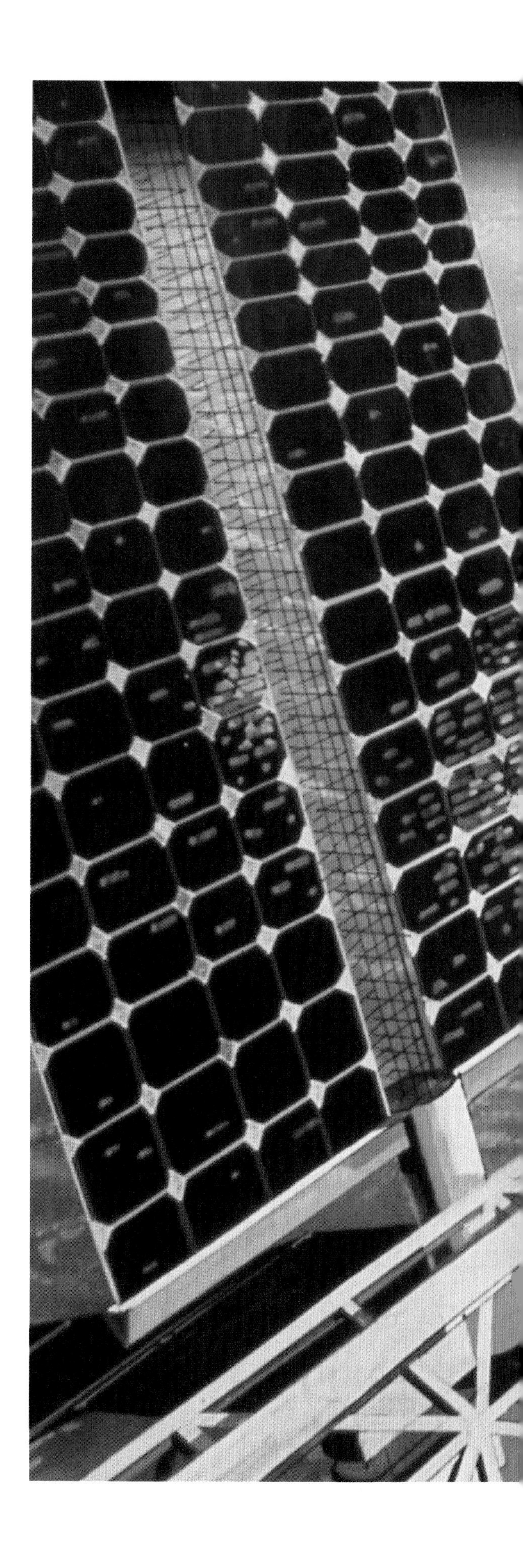

nace, locked in a perpetual heat wave. Until we understand the lessons taught by the history of our neighboring planets, it is dangerous to blunder on as we have in the past, relying on precepts that served us well before. We are up against dynamics we do not yet understand, and what we do not know *can* hurt us.

Columbus sailed west to reach the Indies, an impossible mission with the ships and provisions under his command. His motives were flawed; he imagined that the world was only one-third its true size. His voyage succeeded because he ran into something new, an unknown continent. It would honor his memory were we to discard the myth still taught in many schools—that Columbus "knew" the world was round, and was opposed on this point by scholars in the Spanish court—and to acknowledge that his discovery of the New World was due not to the confident act of an authority figure, but to the bravery of a skilled but erring navigator willing to venture into the unknown. Space exploration is similar: its value resides less in confirming what we already know than in exposing us to something new.

This cause is best served by space probes that act as extensions of our senses—spacecraft like *Viking*, which landed on Mars in 1976 and yielded most of the data we have to this day about the red planet, and *Voyager*, which reconnoitered four giant planets and 57 moons, transmitting back to Earth photographs that vastly expanded our comprehension of the solar system and its evolution. And it can be served by manned missions, too, provided that they are

The open view acknowledges the universe as vast and limitless—a domain for observation, exploration, and learning. The value of space exploration resides less in confirming what we already know than in exposing us to something new.

Overleaf: Royal Observatory, Edinburgh

genuinely exploratory in character. It may make sense to muster the will and wherewithal to send men and women to Mars, if they go to learn from Mars and be changed thereby—to become Martians. But why risk lives and treasure on an expedition whose goals are driven by such paltry pretensions as to plant the flag and strut our stuff and pave the way for Martian mines and shopping malls?

Least well suited to an open, educational imperative are big, costly, low-Earth-orbit projects like the space shuttle (originally sold to Congress as a prerequisite to building a space station) and the space station (which is now being touted because it would give the shuttle something to do). Backed by billions of dollars' worth of power, these missions blast their way into space, chartered less to comprehend than to command. They exemplify what the mathematician and philosopher Jacob Bronowski called black magic, the belief that we can force nature to do our bidding even when we do not properly understand how nature works.* Such an approach can be dangerous on the frontier—as General George Armstrong Custer learned on the last day of his life.

NASA has become a bastion of the Custer mind-set: enchanted with firepower and beholden to a blinkered Congress, it blusters toward allegedly "practical" goals that it assures everyone it can attain. But genuine exploration is always impractical—at first—because its outcome is always unpre-

*Jacob Bronowski, "Black and White Magic," in *Magic, Science, and Civilization* (New York: Columbia University Press, 1978).

dictable: an explorer who knew in advance what he or she was going to find would not be an explorer. We need to acknowledge how little we know, and emphasize the need to learn. Rather than spending billions of dollars on manned low-Earth-orbit missions modeled after trucking companies—pretending that we have "conquered" space, and are now ready to turn it into a parking lot—we could be putting probes into orbit around every planet in the solar system, peppering space with radio telescopes and infrared telescopes and other antennae aimed outward, and generally opening our eyes as wide as possible to the hard facts of how the universe actually behaves.

Ours is a *thinking* species, one that gained dominance of this planet through learning and adaptability, not physical strength. In the heavens, as on Earth, we are better off studying nature than merely trying to push it around. If aliens landed a spaceship on the White House lawn, would they be impressed by our power—our ability to build floating Eiffel Towers in orbit, to aim lethal lasers at foreign cities? Or would they be more impressed to learn that we were at work inquiring into the evolution of stars and galaxies, gauging the dying glow of the big bang, trying to understand how we got here and where we are going? And which set of accomplishments is more likely to benefit our grandchildren and earn their respect?

They are both the same question, really. How we answer it may well decide whether our grandchildren have a shot at the stars.

WALTER E. MASSEY

Discovery and the Art of Science

THE DISCOVERY process by which we explore and expand the bounds of knowledge is a defining characteristic of our humanity. It shows up not only in the grand discoveries of Columbus and Newton, but also in the learning of a toddler and the deliberate experimentation of a scientist.

Discovery, and the process by which one discovery leads to others, can be fully appreciated only with the passage of time. Looking back at the 500 years of discovery since Columbus's first voyage, we can understand how Columbus can be viewed as initiating not only a flood of European exploration, but also a spirit of geographic discovery and conquest. Scientific exploration and discovery has also changed our world enormously, although it lacks the clear starting point of Columbus's voyage. Like the European explorers, scientific exploration and discovery has transformed our world in ways that even after centuries of experience we cannot fully appreciate. How do geographic and scientific discovery compare? What is the nature of modern scientific discovery? What might the future hold?

A LIFETIME OF DISCOVERY

LET US start with a mental exercise that will place history in a more personal frame of reference. I find it helpful to visualize the rich history of discovery in the five centuries since Columbus's first voyage by compressing it into a single lifetime—for example, that of an 80-year-old grandmother.

In this life span, if Columbus's first voyage occurred when this woman was born, the Earth would have been circumnavigated by Magellan's crew before she reached her fifth birthday. St. Augustine, Florida, would have been settled by the time she was 11. She would be 20 when the *Mayflower* landed, 36 when New Orleans was founded, and 52 during Lewis and Clark's exploration of the American West. She would be 59 when the source of the Nile was conclusively determined.

At age 67, she would have noted Amundsen's arrival at the South Pole. The brief time between the Wright brothers' first flight when she was 65 and Lindbergh's transatlantic flight when she was 69 shows the increasing pace of progress in travel, which is also reflected in the 4-year gap between the first jet airplane and the first man on the moon, just after her seventy-sixth birthday.

By the time she reached age 80, those areas of our planet yet to be discovered or visited by humans would be confined to what lies beneath the seas.

In comparison with this lifetime of geographic exploration, consider the evolution of scientific discoveries and technological developments over the same 80-year life. Science as we understand it today would be unknown when our imaginary grandmother was born. And although the seeds of the Copernican revolution would have been planted by her third birthday, alchemy and astrology would remain the principal ways of organizing knowledge during her first two decades.

Naturalist classification would begin in earnest during her teens with the publication of books on mineralogy, anatomy, geography, botany, and zoology. Galileo would have been viewing the heavens through a telescope by the time she reached 17, and she would be 23 in the year that Galileo died and Newton was born.

She would be 41 when Franklin conducted his experiments with electricity, and in her early fifties when the steam engine was developed as a means of land and water transport. From her mid-fifties onward, scientific discoveries would begin to tumble forth at such a rapid pace that it is difficult to select representative illustrations.

Landmark advances in electromagnetism, chemistry, physics, and biology, accompanied by inventions ranging from the telegraph to the sewing machine and by the establishment of the National Academy of Sciences, all occur between her fiftieth and sixtieth birthdays. In her sixty-ninth year alone, protons and electrons are detected, modern rocketry is born, Einstein publishes his theory of general relativity, psychoanalysis is introduced, and the dimensions of the Milky Way are discovered.

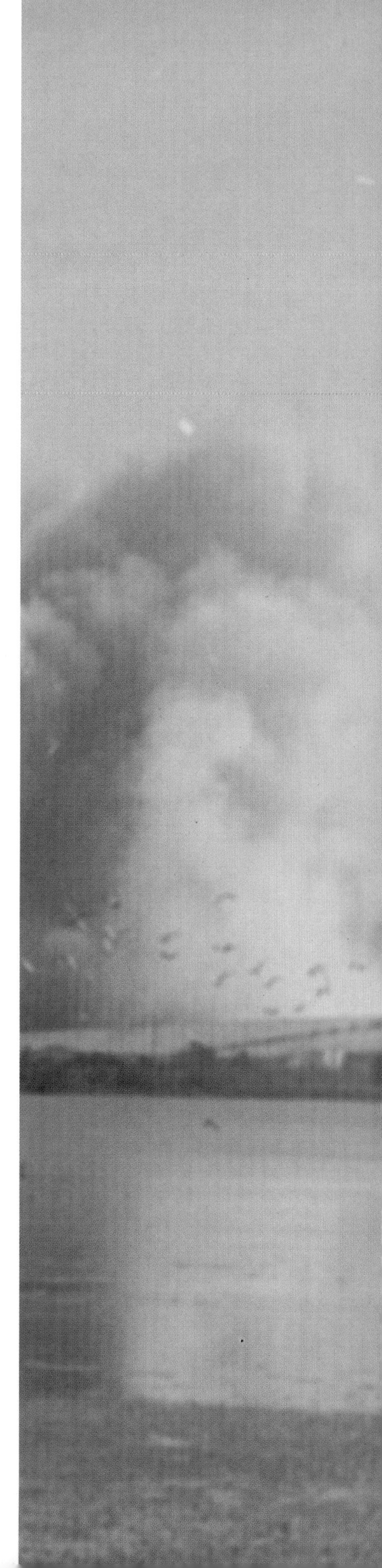

In this century, the link between exploration and technology has tightened. In the span of a single lifetime, technology made possible both the first flight of an airplane and flights to the moon.

Above: Smithsonian Institution; *right:* NASA

What do these abbreviated excursions through time tell us about the relationship between the pace of geographic exploration and that of scientific discovery?

First, geographic exploration by Europeans was not dramatically accelerated by developments in science and technology during the first 300 or so years following Columbus. Sailing ships, horses, and human walking continued to be the principal means of exploration during this period. While there was an undeniable accrual of improvements in navigation and shipbuilding over the centuries, it is difficult to identify any breakthrough near the end of the fifteenth century that markedly increased the chances of a successful transatlantic voyage. Certainly there were important refinements in the magnetic compass during this period, but its use in navigation predated Columbus by as much as 350 years. Similar improvements in techniques to determine latitude more accurately were also coming into use, but they did not represent an advance in scientific understanding or a navigational breakthrough that suddenly made long voyages possible.

In the past century, however, the link has tightened between exploration and technology. Consider, for example, the rapid acceleration of flight technology, which opened remote regions of the world, the atmosphere, and space for exploration. In a single lifetime, it was possible to observe both the first flight of an airplane and a manned flight to the moon.

While today we see science, technology, and exploration advancing in concert, until fairly recently scientific expeditions were often appendages to explorations aimed at discovery, rather than a purpose for them. The Lewis and Clark expedition, set under way by no less an inquisitive mind than that of Thomas Jefferson, had as its primary goal mapping, rather than doing a natural-history survey of, the newly acquired western territories. Even satellite development and space flight were driven as much by tests of technology or potential military applications as by science. The potential of these technologies for scientific discovery was obvious, and planetary exploration and Earth observation have moved them solidly into the scientific realm.

Second, the large-scale European discoveries that began with Columbus are somewhat self-limiting, while scientific and technological advances are self-perpetuating. Once a land mass or a geographic feature is discovered, explored, and described, not much remains for the adventurer but to move on to the next unexplored area. And there are not many of those left on Earth. For the scientist, each new discovery uncovers fertile ground for hundreds of additional discoveries and refinements. And the realm for discovery keeps expanding.

Finally, with the advent of airplanes, terrestrial geographic discovery was in its final throes and its fate was all but sealed when the first surveying satellites began beaming down their images. As the expression goes, we have nowhere to go but up (or down) if we are to continue explorations of discovery. The application of new technologies for scientific discovery, however, continues to open up new realms for examination, from the scale of subatomic particles to that of the universe at large.

BELIEF AND DISCOVERY

GIVEN these contrasting contexts, do voyages of geographic discovery parallel the way that scientists discover the nature of the world around them? What, if anything, connects the contemporary scientific discoverer with the explorer of Columbus's time?

Certainly, the excitement of potential discovery is a common motivating force throughout history. The thrill of being the first to see a new place or a previously unrecognized scientific phenomenon is undeniably compensation for the myriad hardships and sacrifices endured prior to a discovery.

But innate curiosity must be fueled by something more in those who are persistent enough to make important breakthroughs. The common connection between the European explorers and their counterparts in scientific discovery is that their ardent curiosity is sustained by faith. Faith in this context is the confident, enduring belief in the correctness of an idea. Whether that idea reflects spiritual or physical relationships is unimportant. What matters is a commitment to an idea that goes beyond the accepted conventional wisdom—a level of personal conviction that holds doubt at bay.

Columbus and those who followed in his wake had faith in their ability to travel to distant unknown lands and return alive. For the contemporary scientist, this common thread of faith is embodied in the belief that the application of the scientific method will reveal the mysteries of nature and move us to a better understanding of the universe.

Faith in science today is held not only by those who consider themselves to be "doing science," but also by the citizenry at large. This public faith in science is a modern phenomenon. Although support of intellectual pursuits by those who govern is perhaps as old as civilization itself, only in the past 50 years or so have governments provided significant sustained investments in basic scientific research.

This show of faith, backed up as it is by public funds, reflects the expected return on that investment in terms of increased national security and an improved quality of life. But as important as these outcomes are, faith in science is also driven by a belief that even imperfect advances in knowledge and understanding are far preferable to the ignorance and superstition they displace. In this way, faith in science presumes that all increments in knowledge are valuable.

One tenet of our faith in science is that we cannot always predict how or when new scientific knowledge will be applied. The lack of prescience by scientists (and those who apply this new knowledge to technology) about the eventual value of their discoveries necessitates a shared vision that the real benefits of science conducted today may be known only to our children. It is the anticipation of practical value, the successful application of scientifically derived knowledge, that sustains this public support. But how long are we willing to wait before we know if our faith has been rewarded? At what point can we say that science is successful?

For discoverers of new worlds, one measure of success is traveling to a place unknown to their people and—not a trivial consideration—returning alive. If these explorers also return with tales of unparalleled beauty, unfamiliar people and customs, or newfound wealth, so much the better.

For explorers and discoverers of the other type—those who seek new scientific knowledge—the measure of success is much less precise. This imprecision provides an agreeable ambiguity for those eager to understand the next level of the puzzle. But this uncertainty also provides a source of constant tension between the scientist and those who support scientific research.

Like their global-exploring counterparts, scientific discoverers must see or do what no one has seen or done before. But unlike an explorer who can give an immediate report of travels, the scientist's work is unlikely to enter the public mind all at once.

Scientific discoveries and their potential value may be perfectly clear to the scientists working on them, but they typically take a long time to become integrated into public awareness. Often it is years before technology, the more public result of science, develops sufficiently to allow us to recognize the importance of the basic science that preceded it. By that time, the original science may well be lost in the public mind. Nobel prizes are sometimes awarded decades after the original work has been completed. By that time, the science that is being honored has often been applied to a technology that is commonplace. This time lag makes faith in science all the more critical to the survival and growth of the pursuit of knowledge.

Scientific discovery and the discovery that results from the adventure of exploration have some commonalities: both must take us where we have never been before; both must be public and replicable; and both are frequently viewed in the rear-view mirror of history as inevitable.

Yet the scientist is far less likely to be accorded the recognition and public acclaim that is awarded to the explorer. A major reason for this is the inherent caution and skepticism of the scientific method. The scientist operates in a milieu that is both less public than the explorer's and, increasingly, less discernible to the public.

Our faith in science and the technology it spawns is so ingrained that it is difficult to imagine a time when people did not assume that, given an advance in science and technology, any given problem would yield a solution. While we may view the conquest of AIDS or the colonization of the moon as inevitable once scientific and technical problems are overcome, it is unlikely that European sailors in 1450 regarded scientific and technical advances in navigation or shipbuilding as the only things keeping them from safely crossing the Atlantic.

Today we have every confidence that given the time, the resources, and the political will, a manned mission to Mars, for example, is going to happen. We assume that whatever problems exist, scientists and engineers can solve them. The majority of the public have faith that science and technology can make possible the impossible.

We often take public confidence in science for granted. For us, it is natural that people should believe in the rational. But such faith today is no more guaranteed than was the sixteenth-century European view that monarchies were the normal and inevitable form of government. Sustaining public support for scientific exploration and discovery requires an awareness of how this public faith came to be and what threats it faces now and in the future.

THE ART OF SCIENTIFIC DISCOVERY

IF WE ARE to sustain the public's faith in science, we must maintain the integrity of science even in the face of human failings. Ego, self-promotion, pride of authorship, stubbornness, and a host of other human frailties are often magnified in those bold enough to seek new worlds. Ours would be a vastly diminished species without these human characteristics. The art of scientific discovery not only depends on people who exhibit these frailties, but also celebrates the positive good that can result from them.

The art of science requires of its practitioners the ability to see patterns where others may see only jumble. As the artist recognizes when the work is complete, so the practitioners of the art of science must know when to modify the work, when it has been taken as far as it can be taken, and when it is time to scrap one approach and take up a completely new line of inquiry.

Henri Amiel, a nineteenth-century writer, noted that the great artist is a simplifier. The scientist who is successful in the art of discovery is truly a great simplifier. Nothing is more complicated than the unknown, for it masquerades as the unknowable. And nothing is simpler than what we know, for it becomes a part of us.

Scientific discovery is not a perfect art practiced by perfect people, but it leads us to ever clearer visions of the universe. It is the art of discovery that recognizes when to abandon a formerly held position and move on. It is the art of discovery that recognizes that error or failure may harbor the possibility of a larger truth.

Nurturing the public's faith in science is part of the art of science. Maintaining the vision of scientific advances during times of competing needs, providing the historical connections between scientific advances and improvements in our lives, focusing our resources in areas where advances are likely, and reassuring the public that current setbacks are not impossible to overcome are some of the confidence maintenance work that the scientific community must do.

THE FUTURE OF DISCOVERY

EARLIER in this essay, I noted how exploration and discovery in Columbus's time and for centuries afterward occurred virtually independently from advances in science and technology. Today, however, science and exploration are inescapably linked. Whether we envision exploring the next frontier through 8-meter telescopes or through scanning tunnelling microscopes, whether we anticipate a manned mission to Mars, a probe beyond the solar system, or colonization of the ocean floors, we will require the best possible science and engineering to take us there.

The future of exploration will depend not only on synergistic advances in science and technology, but also on expanding our sense of the possible. It is extremely unlikely that Columbus could have imagined a steel ship powered independently of the wind or satellite navigation with an accuracy measured in meters. It is also unlikely that such recent visionaries as the Wright brothers could have foreseen thousands of flights every day, each carrying hundreds of passengers to destinations around the globe.

In spite of the rapid advances in science and technology that we have experienced, our visions of the future are, paradoxically, much less distinct than were those of Columbus. It was natural for Columbus, given the relatively static nature of the world he knew, to assume that the culture he would encounter on the islands off the coast of China would be the same as that reported by Marco Polo almost 200 years earlier.

Today, when we look back at the two centuries since the founding of our country and the myriad transformations that have affected every aspect of our lives, the mind boggles. It is difficult for us to conjure even a blurry image of the society our grandchildren will inherit, much less peer one or two centuries into the future.

Coming generations will think of exploration in a very different way than we have viewed it over the past 500 years. Long and dangerous as Columbus's voyage was, its duration was trivial in the context of space travel. What would the impact of Columbus's voyage have been if it had required 40 years to make the round trip? What if it had required 200 years?

The successful explorer of new worlds in space may have to break the rules and never return home. Would people left behind be willing to put resources into such an adventure, realizing that the only return on their investment would be the slim hope that humanity might expand its presence in the galaxy?

It is not modesty that prevents me from saying that I cannot imagine what people will be writing on the topic of exploration and discovery on the six-hundredth anniversary of Columbus's voyage. Looking into the future, whether it is the future of science or the future of exploration, is both risky and unsettling. Little imagination is required to envision a manned trip to Mars, but imagination strains to envision humans voyaging beyond the planets. The difficulty is not in addressing the scientific or technical problems, but in addressing the social and temporal ones. Even a relatively short 50-year human voyage into space might mean returning to an alien and perhaps unwelcoming Earth. Faced with human limits not of courage or knowledge, or even of imagination, but of the magnitude of time required to sample the 100 billion stars spread out across our own galaxy, we may be forced to focus on future modes of exploration and discovery that are beyond our current ability to imagine.

The challenge of human exploration of space will tantalize us as long as we exist. The expansion of our imagination in the beginning of the era of space travel has provided us with possibilities that would have seemed incredible, even to visionaries, only a few years ago. Advances in astronomy have given us a view of the universe that both emboldens and humbles us. Science offers the discoverers of the future an opportunity to see the universe differently. It expands the world to the limits of imagination, and then expands our imagination to capture this new world. Using science to enhance our ability to discover, both on Earth and beyond, will demand resources, imagination, and, above all, faith.

I wish to acknowledge the very able assistance of David Stonner, Office of Legislative and Public Affairs, National Science Foundation, in the preparation of this essay.

DESTINY: QUESTIONS FOR FUTURE EXPLORERS

DESTINY: the very word conjures up prophecy, laden with import and inevitability. Consider the mantra of our age: our destiny is in space; we are destined to inhabit other worlds and explore among the stars. It is a rhetorically powerful and seductive concept.

Destiny is a future ordained by God or cosmic law, fortune or fate, a will that is not of human intention. It declares optimism and necessity. Yet space exploration is supremely an act of human will and intention. A choice made by some and not by others, it has enriched and ennobled human history, but is it truly inevitable?

These essays about future exploration probe a dilemma: if space exploration is destined, why is it pursued haltingly? Space exploration requires motive, technology, and a sustained commitment of will and resources. It is launched by a compelling social consensus. Carl Sagan eloquently urges the motive, Robert Forward eases the technical challenge (sanguine about the effort involved), and Valerie Neal questions the nature of the requisite commitment.

A latent question about space exploration is whether our ultimate motive is indeed to inhabit other worlds. Curiosity and the quest for knowledge may have been sufficient reason for initial exploration, but the political arena now demands other justifications for returning. Having developed capabilities for space-faring, we should now clarify our motives for continuing the enterprise. Do we explore other worlds in space because we really intend some day to live there?

Hopes are the prelude to plans. At the end of this remarkable century, which has seen people soar away from this world yet also approach the brink of self-destruction, we must be wary of complacent claims of destiny. It is just possible that in choosing a destination, we make our own destiny.

C A R L   S A G A N

Explorers

I KNOW where I was when the space age began. In early October 1957, I was a graduate student at the University of Chicago, working toward a doctorate in planetary astronomy. The previous year, when Mars was the closest it ever gets to Earth, I had been at the McDonald Observatory in Texas, peering through the telescope and trying to understand something of what our neighboring world is like. But there had been dust storms on both planets, and Mars was 40 million miles away. When you are stuck on the surface of the Earth, other worlds are tantalizing but inaccessible.

I was sure that someday space flight would be possible. I knew something about Robert Goddard and V-2 rockets and Project Vanguard and even Soviet pronouncements earlier in the 1950s about their ultimate intentions to explore the planets. But despite all that, *Sputnik 1* caught me by surprise. I had not imagined that the Soviet Union would beat the United States to Earth orbit, and I was startled by the large payload (which, American commentators claimed, must have been reported with a misplaced decimal point). Here the satellite was, beeping away, effortlessly circling the Earth every 90 minutes, and my heart soared—because it meant that we would be going to the planets in my lifetime. The dreams of visionary engineers and writers—Konstantin E. Tsiolkovsky, Robert Goddard, Wernher von Braun, H. G. Wells, Edgar Rice Burroughs—were about to be fulfilled.

Sputnik 1, launched in 1957, was the first artifact of the human species to orbit the Earth. *Mariner 2*, sent off in 1962, was the first spacecraft to explore another planet. These two achievements—one Soviet, the other American—mark a new age of exploration, a new direction for our species: the extension of the human presence to other worlds.

We have always been explorers. It is part of our nature. Since we first evolved a million years or so ago in Africa, we have wandered and explored our way across the planet. There are now humans on every continent—from pole to pole, from Mount Everest to the Dead Sea—on the ocean bottoms, and in residence 200 miles up in the sky.

The first large-scale migration from the Old World to the New happened during the last ice age, around 11,500 years ago, when the growing polar ice caps shallowed the oceans and made it possible to walk on dry land from Siberia to Alaska. A thousand years later, we were in Tierra del Fuego, the southern tip of South America. Long before Columbus, people from Borneo settled Madagascar, off the African coast; Indonesians in outrigger canoes explored the western Pacific; and a great fleet of oceangoing junks from Ming Dynasty China crisscrossed the Indian Ocean, established a base in Zanzibar, rounded the southern tip of Africa, and entered the Atlantic Ocean. In the eighteenth and nineteenth centuries, American and Russian explorers, traders, and settlers were racing west and east across two vast continents to the

Pacific. This exploratory urge has clear survival value. It is not restricted to any one nation or ethnic group. It is an endowment that the human species holds in common.

At just the time when the Earth has become almost entirely explored, other worlds beckon. The nations that pioneered this new age of exploration, the Soviet Union and the United States, were motivated nationalistically, of course, but served as well as the vanguard of our species in space. Their combined achievements are the stuff of legend. We humans have sent robots, then animals, and finally ourselves above the blue skies of Earth into the black interplanetary void. The footprints of 12 of us are scattered across the lunar surface, where they will last another million years. We have flown by some 60 new worlds, many of them discovered in the process. Our ships have set down gently on scorching Venus and chilly Mars, returning images of their surfaces and searching for life. Once above our blanket of air, we have turned our telescopes into the depths of space and back on our small planet to see it as one interconnected and interdependent whole. We have launched artificial moons and artificial planets, and have sent four spacecraft on their way to the stars.

From the standpoint of a century ago, these accomplishments are breathtaking. From a longer perspective, they are mythic. If we manage to avoid self-destruction, so that there *are* future historians, our time will be remem-

From the standpoint of a century ago, the accomplishments of space exploration are breathtaking; from a longer perspective, they are mythic.

Right: NASA; *overleaf left:* Olympus Mons, Mars; *right:* Uranus, as seen from its moon Miranda (NASA)

bered in part because it was when we first set sail for other worlds. In the long run, as we straighten things out down here, there will be more of us up there. There will be robot emissaries and human outposts throughout the solar system. We will become a multiplanet species.

We are not motivated by gold or spices or slaves or a passion to convert the heathen to the One True Faith, as were the European explorers of the fifteenth and sixteenth centuries. Our goals include exploration, science and technology, national prestige, and a recognition that the future is calling. There is a very practical reason as well. We can take better care of the Earth (and its inhabitants) by studying it from space and by comparing it with other worlds.

But whatever our reasons, we are on our way. We advance by fits and starts; there are detours and failures of nerve. The long-term trend, though, is clear: it is getting cheaper and easier to go into space, and there is progressively more for us to do there.

Only a handful of nations have access to space at the moment, but their number is increasing. France and China are now lifting commercial payloads for a profit. Japan and the European Space Agency, in 1986, mustered their first, extremely successful missions into interplanetary space. There will be other space-faring nations in the next few decades. Others may lose their determination and their vision, as did Portugal, which trail-

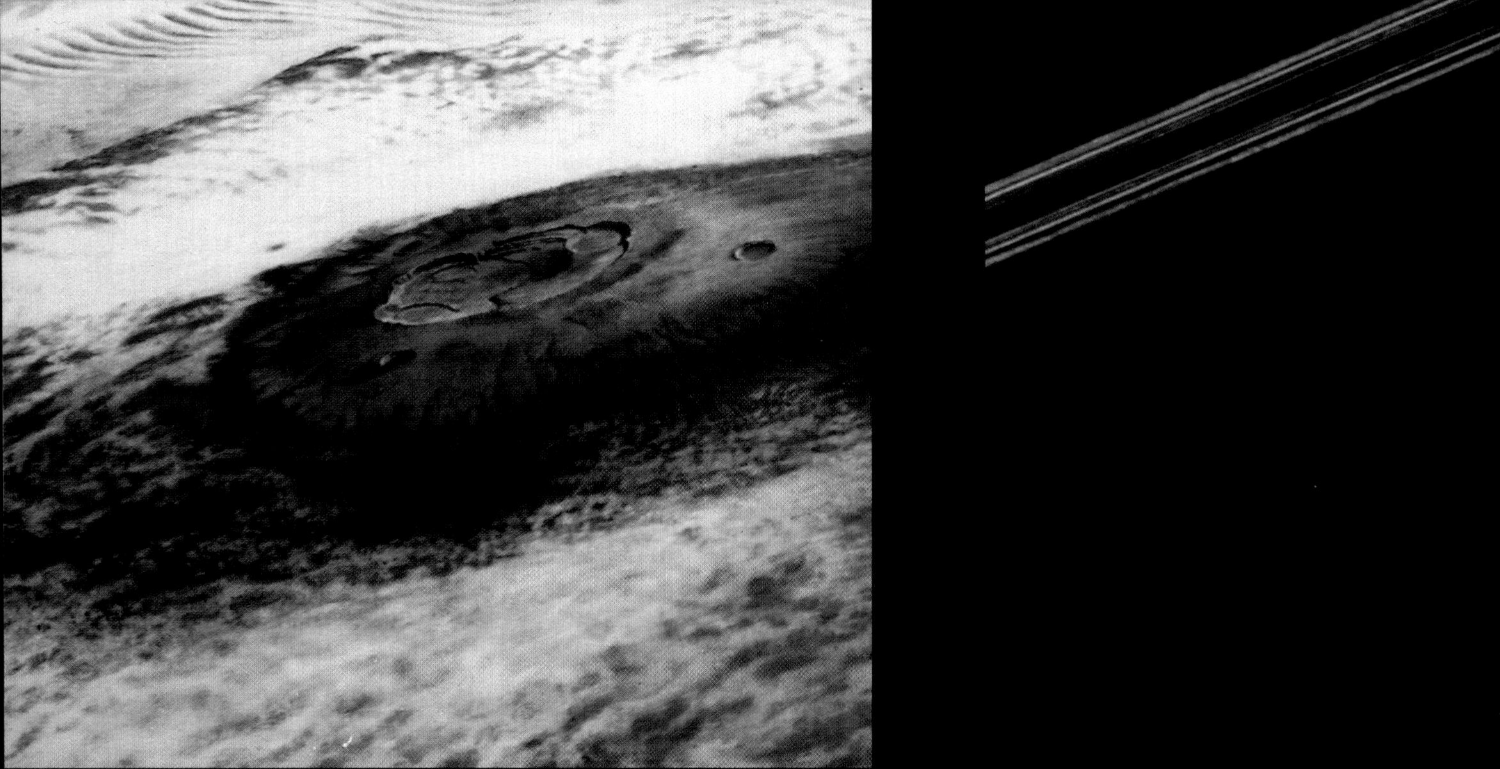

As we become a multiplanet species, there will be robot emissaries and human outposts throughout the solar system.

Pat Rawlings, *Mars Base* (1985)

blazed the great sailing-ship voyages of discovery and then gradually sank into obscurity.

With the competitive impetus of the cold war over, the Soviet Union disintegrated, and the United States facing grave fiscal and other domestic problems, it is by no means clear that those nations that pioneered the early exploration of space will be engaged in the enterprise when its real fruits begin to be harvested. But this need not be the case, and with a real dedication to international cooperation—saving costs and pooling resources—the human species is fully able to continue its exploring.

It has been my good fortune to have participated, from the beginning, in this new age of exploration; to have worked with those glistening *Mariner*s, *Apollo*s, *Pioneer*s, *Viking*s, and *Voyager*s in their journeys between the worlds, a technology that harmed no one and that even America's adversaries admired and respected; to have played some part in the preliminary reconnaissance of the solar system in which we live. I feel the same joy today in these exploratory triumphs that I did when *Sputnik 1* first circumnavigated the Earth, when our expectations of what technology could do for us were nearly boundless.

But since that time, something has soured. The anticipation of progress has been supplanted by a foreboding of technological ruin. I look into my daughter's eyes and ask myself what kind of future we are preparing for our

NASA
SERVICES

children. We have offered them visions of a future in which—unable to read, to think, to invent, to compete, to make things work, to anticipate events—our nation sinks into lethargy and economic decay; in which ignorance and greed conspire to destroy the air, the water, the soil, and the climate; in which we permit a nuclear holocaust. The visions we present to our children shape the future. It *matters* what those visions are. Often they become self-fulfilling prophecies. Dreams are maps.

I do not think it irresponsible to portray even the direst futures; if we are to avoid them, we must understand that they are possible. But where are the alternatives? Where are the dreams that motivate and inspire? Where are the visions of hopeful futures, of times when technology is a tool for human well-being and not a gun on hair trigger pointed at our heads? Our children long

The visions we present to our children shape the future. It matters what those visions are.

MariLynn Flynn, . . . *And Don't Slam the Airlock!* (1989)

for realistic maps of a future that they (and we) can be proud of. Where are the cartographers of human purpose?

Continuing, cooperative planetary exploration cannot solve all our problems. It is merely one component of a solution. But it is practical, readily understood, cost-effective, peaceful, and stirring. It is our responsibility, I believe, to create a future worthy of our children, to fulfill the promise made decades ago by *Sputnik 1* and *Mariner 2*, to open up the solar system to those intrepid explorers from planet Earth.

ROBERT L. FORWARD

Technological Limits to Space Exploration

Can We Sail to the Stars?

THERE is only one technological limit to space exploration. It is a speed limit—the ultimate speed limit. Nothing can travel faster than the speed of light (300,000 kilometers, or 186,000 miles, per second). There are no other technological limitations.

Certainly, there exist economic, environmental, and Earth-resource limitations, and, always, the limitations of the social consensus that arise from lack of will, failure of nerve, or unwillingness to take risks. But there is only one technological limitation, and even that cannot keep some future Columbus from sailing to the stars.

It is difficult to go to the stars. They are far away, and the speed of light limits us to a slow crawl along the star lanes. Decades and centuries will pass before the stay-at-homes learn what the explorers have found. The energies required to launch a manned interstellar vehicle are enormous, for the mass to be accelerated is large and the cruise speed must be high. Yet even these energies are not out of the question once we move our technology out into space, where the constantly flowing sunlight is a never-ending source of energy. There are many ideas in the technical literature on methods for achieving interstellar transport. In time, one or more of these dreams will be translated into a real starship.

But be warned. Many people (some of them quite well known) have "proved" by "calculation" that interstellar flight is "impossible." Actually, in each case, all that those people "proved" was that with their choice of "obvious" assumptions, interstellar travel was made so difficult that they were unwilling to consider it further. Some examples of these "obvious" assumptions are that a self-contained rocket has to be used, that the rocket has to travel at a constant acceleration of 1 g, that the mission has to be completed in 10 years, and that all the rocket fuel has to be extracted from Earth resources.

It is *not* feasible to carry out rapid interstellar travel with conventional rocket technology. If rockets are used to propel a space vehicle, even if the rockets use nuclear fuel, the vehicle will be limited in its terminal velocity to a small fraction of the speed of light. If a rocket-powered spacecraft has a human crew, the spacecraft will have to be designed as a "world ship," in which the crew lives for many decades or even generations during the long journey between the stars. To get to the stars in less than a human lifetime, interstellar vehicles must use some form of "rocketless rocketry," by which the vehicle does not carry along its energy source and/or reaction mass and/or some other parts of the conventional rocket.

It is *not* feasible to carry out interstellar travel at a constant acceleration of 1 g. After the first year of acceleration, the vehicle is moving at 70 percent of the speed of light. From then on, the energy used to supply propulsive thrust does not make the vehicle go significantly faster (to the people at home who are paying for the mission). Instead, all that energy just goes into making the vehicle heavier and harder to push. A properly optimized interstellar mission accelerates up to some cruise velocity well below the speed of light and then coasts the rest of the way, cutting energy and fuel requirements by orders of magnitude.

It is *not* feasible to carry out interstellar travel with a round-trip time of 10 years. Even light requires 8.6 years to get to the nearest star system and back. If we grant that interstellar missions are going to require trip times of 30 to 50 years or longer, the cruise velocity needed to carry out a mission to the nearer stars drops from near the speed of light to less than half that. This reduces many problems caused by the rapid motion of the spacecraft through the gas and dust in the interstellar medium, such as erosion damage to the spacecraft and radiation damage to the crew.

It is *not* feasible to carry out interstellar travel using only the resources of Earth. Although interstellar vehicles can be built without straining Earth resources, the reaction mass, and especially the energy required to drive the vehicles, should be extracted from space.

It is *not* feasible to carry out interstellar travel if one insists on those "obvious" assumptions. Yet, as we shall see, interstellar travel *is* feasible if the proper assumptions are made and the proper technologies are used.

Other people might admit the ultimate physical feasibility of interstellar flight, but question the desirability of sending out spacecraft when we can gain information about interstellar intelligent beings more cheaply and much faster by listening for their messages with radio telescopes on Earth. Whether SETI (Search for Extraterrestrial Intelligence) is really cheaper and faster than CETI (Communication with Extraterrestrial Intelligence) by interstellar probes is debatable. It is also easy to dream up alien civilizations that are intelligent, have information and technology that would be of value to us, and yet, because of their environment (an underwater civilization, for example), do not and will not have radio technology.

If there are beings with radio out there, *if* they are willing to transmit messages into the unresponsive void (instead of just listening as we are), and *if* we listen in the right direction at the right time and the right frequency with the right bandwidth and the right detection scheme, then the radio approach to SETI will make a significant contribution to our knowledge. However, interstellar exploration with automated probes, although still decades in the future, is definitely more certain to produce a contribution of equivalent value. As the famous science fiction writer Arthur C. Clarke said in 1968: "This proxy exploration of the Universe is certainly one way in which it would be possible to gain knowledge of star systems which lack garrulous, radio-equipped inhabitants."*

* Arthur C. Clarke, *The Promise of Space* (New York: Harper & Row, 1968).

It is difficult—but not impossible—to go to the stars. They are far away and the speed of light limits us to a slow crawl, but there are a number of technically possible ways to travel in our galactic neighborhood.

Overleaf: Jon Lomberg and the National Air and Space Museum, *Portrait of the Milky Way* (1992)

INTERSTELLAR DISTANCES

IT IS DIFFICULT to comprehend the distances involved in interstellar travel. Of the billions of people living today on this globe, many have never traveled more than 40 kilometers (24 miles) from their place of birth. Of these billions, only 12 have traveled as far as the moon, which at a distance of almost 400,000 kilometers is 10,000 times 40 kilometers away. In 1989, one of our interplanetary probes, *Voyager 2*, passed by the orbit of Neptune, which at slightly more than 4 billion kilometers is 10,000 times farther out than the moon. However, the nearest star, at 40 trillion kilometers, is 10,000 times farther than Neptune.

Distances this large can best be measured in light-years. A light-year is the distance that light travels in 1 year at 300,000 kilometers per second. It takes 4.3 years for light to travel from the sun to the nearest star, Proxima Centauri, part of a three-star system called Alpha Centauri, in the southern constellation Centaurus. (One of the stars, Alpha Centauri A, is similar to our sun, while Alpha Centauri B and Proxima Centauri are small red-dwarf stars.) To carry out even a one-way robotic probe mission to the nearest star system in the lifetime of the humans who launched the probe will require a minimum spacecraft velocity of 10 percent of the speed of light. At that speed, it will take the automated probe 43 years to get there and 4.3 years for the information to be radioed back to us.

Farther away, in other parts of the sky, are other stars similar to our sun that are our best candidates for finding an Earth-like planet. These sun-like single-star systems are Epsilon Eridani, at 11 light-years, and Tau Ceti, at 12 light-years. To reach these stars in a reasonable time will require spacecraft velocities of 30 percent of the speed of light or greater. At 30 percent of the speed of light, it will take nearly 40 years for a space probe to get to the star, plus another 11 to 12 years for the radio signals to return to Earth.

Although we need to exceed 10 percent of the speed of light to get to *any* star in a reasonable time, if we can attain a cruise velocity of 30 percent of the speed of light, then there are 17 star systems with 25 visible stars and hundreds of potential planets within 12 light-years' distance. This many stars and planets within reach at 30 percent light speed should keep us busy exploring while our engineers are working on designs for faster starships.

What kind of starships can we envision? It turns out there are many, each using a different technology. The three discussed here use nuclear-pulse propulsion, antimatter propulsion, and beamed-power propulsion. For these technologies, we know the basic physical principles and have demonstrated the ability to achieve the desired reactions on a small scale. All that is needed for the design, engineering, construction, and launch of the starship is the application of large amounts of money, manpower, material, and energy—lots of energy. The most powerful source of energy at our immediate disposal is nuclear energy.

NUCLEAR-PULSE PROPULSION

THE OLDEST design for an interstellar spaceship is one that is propelled by nuclear bombs. Called the *Orion* spacecraft, it was invented in the late 1950s at Los Alamos National Laboratory. The original goal was to send manned spacecraft to Mars by 1968 at a fraction of the cost of the Apollo program.

The *Orion* works by ejecting a small fusion bomb out the rear of the vehicle, where the bomblet explodes. The debris from the explosion strikes a "pusher plate," which absorbs the impulse from the explosion and transfers it through large "shock absorbers" to the body of the spacecraft. Although it seems impossible that anything can survive a nuclear explosion, a properly designed pusher plate made of strong, high-temperature materials can survive not only one nuclear-bomblet explosion, but many explosions (provided the bomblets are not too powerful, and the pusher plate is not too close to the body of the spacecraft and is designed to temporarily give ground when impacted by the explosive debris).

The physicist Freeman Dyson took these ideas for an interplanetary spacecraft and extrapolated them to an interstellar spacecraft. The ship would necessarily be large, with a payload of some 20,000 metric tons (enough to support many hundreds of crew members). The total mass would be 400,000 tons, including a fuel supply of 300,000 fusion bombs weighing

about 1 ton each. (This would use up the world's present stock of nuclear-bomb material, and I cannot think of a better way to dispose of it!) The bombs would be exploded once every 3 seconds, accelerating the spacecraft at 1 g for 10 days to reach a velocity of 3 percent of the speed of light. At this speed, *Orion* would reach the nearby Alpha Centauri star system in 130 years. To have a deceleration capability at the target star, this ship would have to be redesigned to have two stages, with the first stage weighing 1.6 million tons.

Although the *Orion* spacecraft has a minimal performance for a starship, it is one form of interstellar transport that could have been built and sent on its way in the last decade. If there developed an urgent need to send a small fraction of the human race to the stars (for example, if we knew that the sun was going to explode in 10 years), then this existing technology could be used to build an interstellar ark.

Orion *was an early, innovative concept for a spaceship propelled by small atomic bombs. Although a rather inefficient transportation method, nuclear-pulse propulsion was based on existing technology.*
Rick Sternbach

The reason for the relatively poor performance of *Orion* is that it uses the rocket principle: the vehicle must carry its own fuel. Under this condition, the final velocity of the vehicle depends on the energy release from the fuel. Since a fusion reaction converts less than 1 percent of its mass to energy, the final velocity of the vehicle is limited to 3 percent of the speed of light. If we wish to use the rocket principle for interstellar transport, we will need a nuclear fuel that has a higher energy-conversion efficiency than fusion, such as antimatter.

ANTIMATTER PROPULSION

ANTIMATTER is a type of matter that has an electrical charge opposite to the electrical charge on normal matter. Because the charges on an antimatter particle and a matter particle are opposite, the antimatter and matter attract each other. Amazingly enough, when the two particles contact each other, not only do their charges equal out and disappear, but their masses disappear in a process called annihilation, in which the mass of the two particles is transformed into enormous amounts of energy.

A spacecraft using antimatter as its source of propulsion energy could "drive" anywhere in the solar system in a week and travel to the nearest stars in a human lifetime. The antimatter should be in the form of antiprotons, since upon annihilation they generate high-energy, electrically charged particles. Because the particles have an electrical charge, they can be contained and directed by magnetic fields to provide thrust. Also, instead of using equal parts of antimatter and normal matter, it is best to use a small amount of antimatter to heat tons of normal matter. For example, to accelerate a 1-metric-ton robotic probe to 10 percent of the speed of light requires only 4 tons of liquid hydrogen for propellant and 9 kilograms of antimatter for the energy source.

Most people do not realize that antimatter is being made daily in laboratories at CERN in Switzerland and Fermilab in the United States. Researchers have also captured antimatter particles and kept them in a magnetic "bottle" for over a month without losing any. In the coming decades, we will see the production and storage of significant quantities of antimatter. If no other propulsion system proves to be better, and if we wish to spend the time and money to generate kilograms of antimatter, then one of these days we can ride to the stars on a jet of annihilated matter and antimatter.

BEAMED-POWER PROPULSION

ALTHOUGH an antimatter rocket is the ultimate in rockets, it is not necessary to use the rocket principle to build a starship. A rocket consists of payload, structure, reaction mass, and energy source. (In most rockets, the reaction mass and energy source are combined into the chemical or nuclear "fuel.") Because a rocket has to carry its fuel with it, its performance is significantly limited. It is conceptually possible to build a spacecraft that does not have to carry along any fuel and that consists only of payload and structure. Two versions of such spacecraft that could be built with "reasonable" extrapolations of present-day technology are a microwave-beam-pushed wire-mesh probe and a laser-beam-pushed lightsail. Both these concepts use the physical principle that a beam of electromagnetic radiation bouncing off a reflector (microwaves off a metal screen or light off a mirror) gives the reflector a tiny, but significant push. These "radiation pressure" forces are well known to space physicists and are already being used by spacecraft engineers to help orient spacecraft in space.

Beamed-power-propulsion systems require lots of energy. Although it is conceivable that the energy can be obtained from some sort of fuel, the fuel will be expensive. Out in space there exists a form of energy that is free. It is the light from our sun—thousands of watts per square meter, billions of watts per square kilometer—constantly pouring out into empty space and unused. Unfortunately, the light from the sun gets weaker and weaker the farther from the sun we travel. To use sunlight to travel to the stars, we need to learn to gather the sunlight and convert it to a more manageable form of light that we can beam to the stars. Once we have learned to harness and redirect sunlight, then it will be possible for some future Columbus to literally *sail* across

To go to the stars in less than a human lifetime requires some form of "rocketless rocketry," a spacecraft that carries no fuel. Starwisp *is a concept for a beamed-power-propulsion vehicle "pushed" by microwave radiation. Freed from the burden of onboard fuel,* Starwisp *could accelerate to the high speed necessary for interstellar travel.*

Rick Sternbach

the interstellar gulf to the small groups of island planets expected to be found orbiting around the nearest stars.

Microwave-Pushed *Starwisp*

Starwisp is a concept for a lightweight, high-speed interstellar flyby robotic probe pushed by the radiation pressure from a beam of microwaves. The basic structure of *Starwisp* is a large sail made of fine wire mesh, with microcircuits at each wire intersection. Each microcircuit would contain a simple photodetector and a complex logic circuit. The sail would be pushed at high acceleration using a microwave beam produced by a solar-power satellite out in space and formed into a long-distance beam by a large concentric-ring transmitter lens made of alternating wire-mesh rings and empty rings. The high acceleration would allow *Starwisp* to reach a coast velocity near that of light while still in the solar system.

Upon the arrival of *Starwisp* at the target star, the microwave transmitter back in the solar system would flood the target-star system with microwave power. Using the wires in the *Starwisp* mesh as microwave antennas, the microcircuits could collect enough power to energize their photodetectors and logic circuits. The collection of interconnected microcircuits would be designed to act as a semi-intelligent neural-net computer with photodetector information as its input and microwave transmitters as its output. In essence, *Starwisp* would

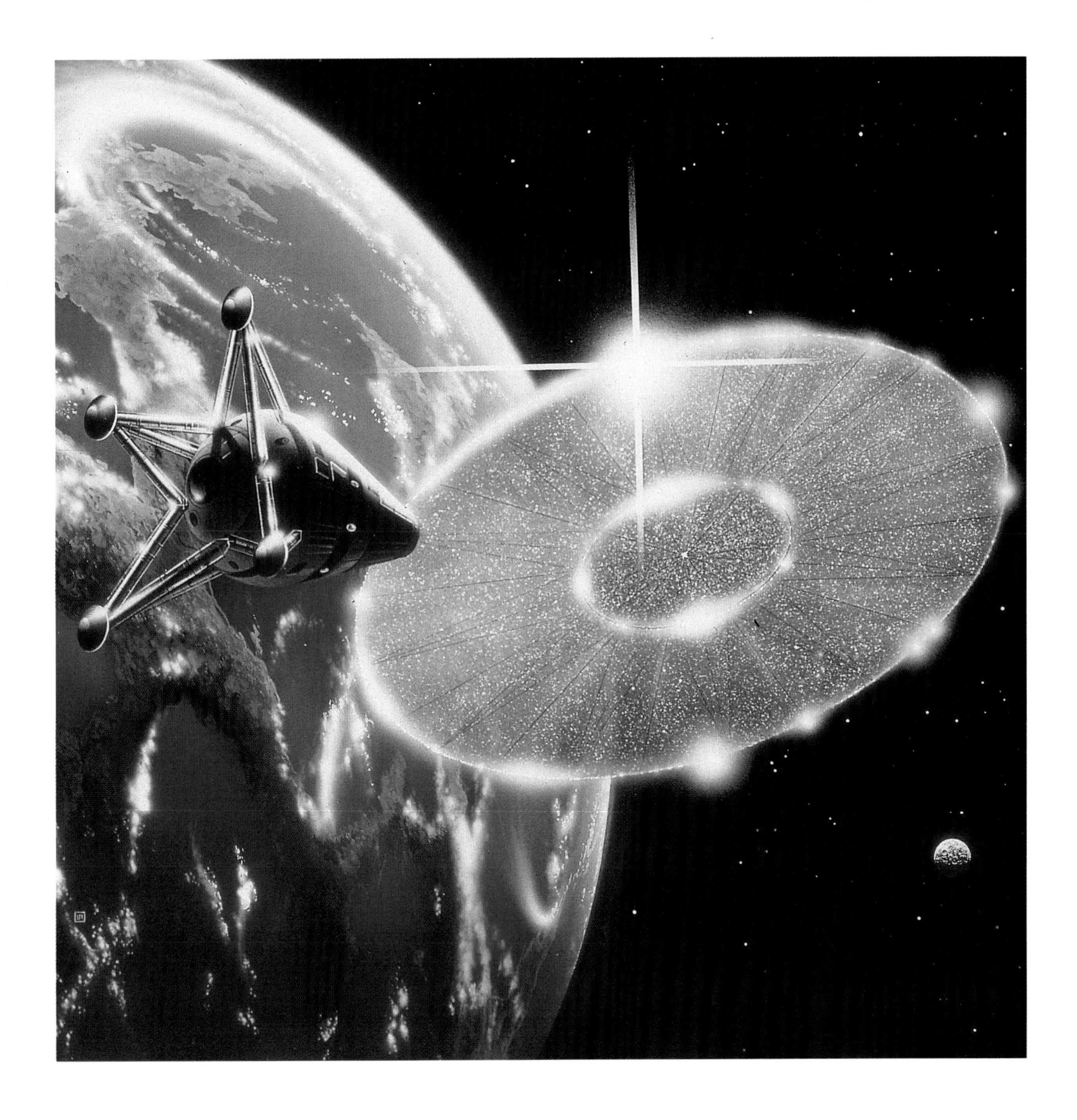

act as a giant artificial retina that would collect the light from the target star and any planets orbiting it, and use the computational capability of the micro-circuits to construct detailed images of the planets. At the same time optical information is being gathered, the microcircuits would measure the phase of the microwaves in the power beam from the solar system. This phase information would be used by the computer to configure the microwave-transmitter outputs so that they would beam the picture-containing microwave signals directly back toward Earth.

A minimal-weight *Starwisp* design would be a 1-kilometer mesh sail weighing 16 grams and carrying 4 grams of microcircuits. (This is a total of 20 grams—less than 1 ounce!) *Starwisp* would be accelerated at 115 g's by a 10,000-megawatt microwave beam, reaching 20 percent of the speed of light in a few days. Upon arrival at Alpha Centauri 21 years later, *Starwisp* would collect enough microwave power to send back real-time, high-resolution color-television pictures during its fly-through of the Alpha Centauri star system.

Because of *Starwisp*'s very small mass, the beamed-power level needed to drive a minimal *Starwisp* is about that planned for the microwave-power output of a solar-power satellite. Thus if power satellites are constructed in the next few decades, they could be used to launch one or more *Starwisp* probes to the nearer stars during their "checkout" phase. Once the probes have found interesting planets, then we can use another form of beamed-power propulsion to visit them in person. Although microwave beams can be used only to "push" a spacecraft in a direction away from the solar system because the microwave beam spreads rapidly with distance, if we go to laser wavelengths where the spread with distance is lower, then it is possible to design a laser-propulsion system that can use a laser beam sent out from the solar system to push the spacecraft to a star and back again.

Laser-Pushed Lightsails

This method for traveling to the stars would use large sails of light-reflecting material pushed by the light pressure from a large laser array in orbit around the sun. With this technique, we could build a manned spacecraft that not only could travel at a reasonable speed to the nearest stars, but also could stop, allow time for exploration, and then return its crew back to Earth, within the lifetime of the crew. It will be some time before our engineering capabilities in space will allow us to travel to Mercury and build the laser system needed, but there is no new physics involved, just a large-scale engineering extrapolation of known technologies recently tested as part of the Strategic Defense Initiative program.

The lasers would be in orbit around Mercury, which, in turn, is in orbit around the sun. Since the light pressure from the sunlight collected by the lasers would have a tendency to push them away from the sun, and the laser light emitted by the lasers would have a tendency to push them in the direction opposite to that of the laser beam, by putting the lasers in orbit around Mercury, the gravity pull of Mercury would keep them in place at a constant distance from the sun.

The lasers would use the abundant sunlight at Mercury's orbit to produce coherent laser light, which would be collected into a single coherent beam and sent out to a transmitter lens floating between Saturn and Uranus. The transmitter lens would be a concentric-ring lens designed to operate at the wavelength of the laser. The lens frame would consist of wires arranged in a spiderweb pattern, kept rigid by a slow rotation about its axis. On the frame would be 1000 ring-shaped regions, alternately empty or covered with a 1-micron-thick plastic film. The lens would be 1000 kilometers in diameter and weigh about 560,000 metric tons. A lens this size could send a beam of laser light over 40 light-years before the beam would start to spread. This would ensure that *no* laser light would be lost due to spreading of the light beam for the entire duration of the mission, and all the laser light would fall on the sail of the spacecraft.

For a mission to Epsilon Eridani at 11 light-years' distance, the lightsail should be 1000 kilometers in diameter (as big as Texas) and made of thin aluminum film stretched over a supporting structure. The total weight would be 80,000 tons, including 3000 tons for the crew, their habitat, their supplies, and their exploration vehicles. The lightsail would be made in three sections, with a circular central payload section 100 kilometers in diameter, surrounded by a middle, ring-shaped section 320 kilometers in diameter (as big as Connecticut) with a 100-kilometer-diameter hole, and an outer, ring-shaped section 1000 kilometers in diameter (as large as Texas) with a 320-kilometer-diameter hole. As will be shown, by making the lightsail in three sections, the lightsail

In this century, in less than one generation, people developed the means for orbital and planetary spaceflight. Might interstellar flight likewise be realized someday?

Overleaf: Andrei Sokolov, *Man Made Stars* (1975) (National Air and Space Museum)

could be reconfigured during the mission to allow the crew to stop at the target star system and later sail back home to the solar system.

The three-stage lightsail would be accelerated at 30 percent of Earth gravity by 43,000 terawatts of laser power. (One terawatt equals 1 trillion watts. For comparison, the entire Earth presently produces only about 1 terawatt of electricity. The laser-power level required is one of the reasons why the laser could not be situated on Earth and powered by Earth resources. Instead, it would be near the sun, where there is plenty of free energy.) At an acceleration of 30 percent of Earth gravity, the lightsail would reach a velocity of half the speed of light in 1.6 years, while traveling 0.4 light-year's distance. The expedition would now coast for 10 light-years, reaching the Epsilon Eridani star system after 22 years (Earth time), or 19 years (crew time).

At 0.4 light-year from the star, the rendezvous portion of the lightsail, consisting of the circular central payload section plus the middle-ring section, would be detached from the center of the lightsail and turned to face the large outer-ring sail. The laser light from the solar system would reflect from the outer-ring sail, which would act as a retro-directive mirror, sending the laser light back toward the solar system. The reflected light from the outer-ring sail would hit the smaller rendezvous sail and decelerate it to a halt in the Epsilon Eridani system. The outer-ring sail, its task completed, would continue on into interstellar space.

1975 A. C

After the crew explores the planetary system around Epsilon Eridani (using their lightsail to travel from planet to planet in the light from Epsilon Eridani), it would be time to bring them back to Earth. To do this, the circular central payload section would be detached from the center of the middle-ring-section and turned to face it. Provided that someone back in the solar system remembered to turn on the laser 11 years earlier, the laser light from the solar system would hit the ring-shaped remainder of the rendezvous sail and be reflected back on the payload sail, sending it on its way back to the solar system. When the payload sail approached the solar system 20 Earth-years later, it would be brought to a halt by a final burst of laser power. The members of the crew would have been away for 51 years (including 5 years of exploring), would have aged 46 years, and would be ready to retire and write their memoirs.

FUTURE PROSPECTS

THERE seems to be only one technological limitation to space exploration—the speed of light. But that limitation still leaves some future Columbus with a lot of territory to explore. It will be difficult to go to the stars, but not impossible. There are not one, but a number of technologies—such as antimatter, solar-power satellites, and high-power lasers—under intensive development for other purposes that, if suitably modified and redirected, can give the human race a flight system that will reach the nearest stars. All it really takes is for the social consensus to have the desire and the commitment to a few decades of difficult and expensive space-engineering work, and our first interstellar probe could be sailing to the stars within our lifetimes. Failing that, now that war machines are not draining our pockets, perhaps one of these days the world will be rich enough that we can afford to do it just as a lark.

VALERIE NEAL

Where Next, Columbus?

EVERY day, going to and from my office in the National Air and Space Museum, I pass the bronze plaque that dedicates the museum to all explorers of air and space. I am reminded that the museum is not about just aircraft and spacecraft—the technology of flight—but about something more transcendent: humans leaving the surface of this world on the wings of a desire to explore.

We have become space-farers, but only tentatively. We stepped off our home planet, briefly visited the moon, and then came back home. We landed a pair of robots on Mars, said thank you very much for the images and data, and did not send another explorer to that tantalizing world until 16 years later. We sent two *Voyager* spacecraft past the giant outer planets, admired the snapshots, and waited more than a decade before launching another explorer for a closer look.

We *can* explore worlds beyond our own, but we are taking our time about it. As a nation, as the human race, we are not yet sure that it is something we really want to do. We hear inspirational talk about our destiny, but we cannot assume that we have a future in space. Why? Because exploration does not just happen; it is an enterprise based on values and choices and commitment. As decisions about the next 50 years—much less the next 500—have not been made yet, Where next? is an open question.

What are some of the issues and challenges we face as a space-faring peo-

ple? Why is it that we have not made up our minds whether to continue moving out to new worlds in space? What else should be considered in the public dialogue about our future in space?

IMAGES AND AMBIVALENCE

IT IS a cliché that we Americans are a nation of explorers, but how do we think about space exploration? What images and metaphors come to mind? In the past 50 years, three concepts of exploration have predominated, and a fourth is now on the rise. The words and images that evoke space exploration help to shape our attitudes and perhaps contain clues to our ambivalence about the future.

A powerful concept is that it is human nature to explore; we are genetically and biologically programmed to be on the move, and we explore in

The National Air and Space Museum is dedicated not to the technology of flight, but to something more transcendent: people leaving the surface of this world on the wings of a desire to explore.

National Air and Space Museum

response to some primal need. This concept of exploration is linked to the concept of conquest, of spreading out to acquire more land, more food, more wealth, more power. In this concept, exploration is an inevitable "next step." While most human cultures do explore and migrate, there are some that inexplicably do not, thus casting doubt on the notion of biologically programmed impulse. We are learning more about the darker side of conquest. While benefits accrue to the explorer, the explored land or people, even another planet, usually does not fare as well. It is often a short step from exploration to exploitation, and the two terms are used interchangeably in space-program rationales. We heard more about the "conquest of space" in the cold war era, but the phrase still resonates in space-policy studies and speeches.

The second concept is that it is our destiny to explore, that exploration is in our national character. This idea of exploration is wed to the West and pioneering. Space is the "high frontier," and it is manifest destiny that people, especially Americans, go pioneering there. At least one historian has pointed out that the space program has adopted only parts of the pioneering analogy—the parts about courage, self-reliance, ingenuity, and taming the wild. There are also valuable lessons to be learned in what went wrong in the westward expansion, but they are usually ignored when the analogy is made to the "new frontier in space."

Nevertheless, pioneering seems to be the prevailing metaphor for future

We have become spacefarers,
but only tentatively. The last person
left the moon in 1972.

NASA

exploration. Space exploration was an essential part of President Kennedy's New Frontier agenda. In 1969, the Space Task Group's recommendations for the future opened with this vision: "We see a major role for this nation in proceeding from the initial opening of this frontier to its exploitation for the benefit of mankind, and ultimately to the opening of new regions of space to access by man."* About twenty years later, the image was reinvigorated when the National Commission on Space (Paine Commission) heralded the future with a report titled *Pioneering the Space Frontier*.† Almost every study or speech about space exploration still echoes this phrase.

The third concept is that exploration is a badge of character, the mark of leadership and superiority. Nations that cease exploring lose their vigor, fall behind, and suffer defeat in the arena of international power and prestige. This concept is linked to competitiveness in economic strength, advanced technology, higher education, and standard of living. Competitiveness is a main theme in the 1987 report *Leadership and America's Future in Space* (Sally Ride report), in the 1988 Presidential Directive on National Space Policy, and in the 1991 synthesis group (Stafford Committee) report on America's Space

---

* *The Post-Apollo Space Program: Directions for the Future*, Space Task Group Report to the President (September 1969), i.

† *Pioneering the Space Frontier: The Report of the National Commission on Space* (New York: Bantam Books, 1986).

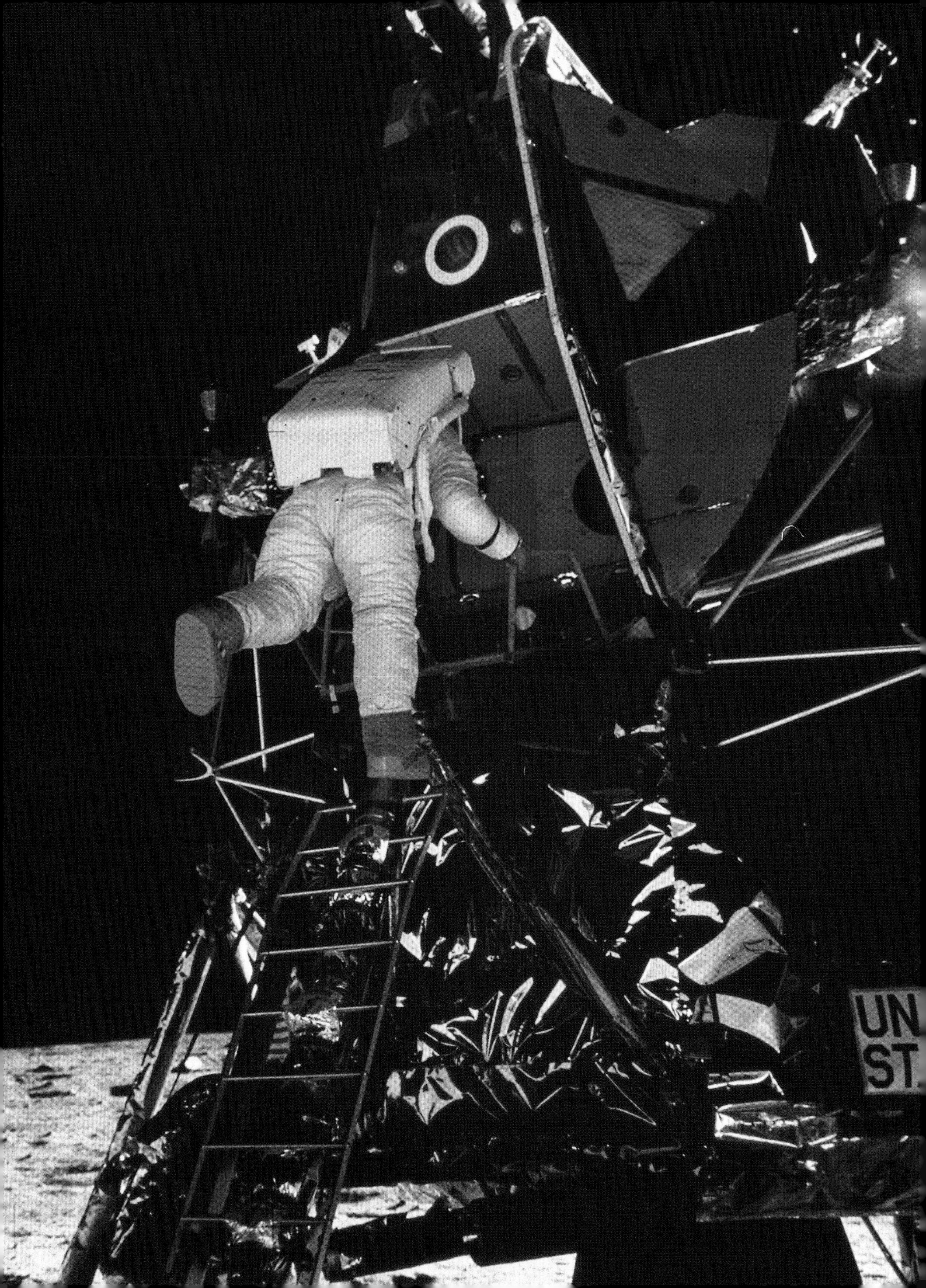
UN
ST.

We admire the spectacular images of other worlds and mine the data from previous robotic explorers, but do we take for granted the future of space exploration?

NASA

Exploration Initiative.* The principal idea is that to remain competitive and maintain world leadership, the United States must actively explore space. If we do not do it, someone else will, and we will be second best.

The conquest metaphor no longer seems valid—what is there to conquer in space?—and the frontier metaphor is beginning to sound exhausted. Competitiveness seems to be the reigning theme in the political and economic arena today; as if the United States does not have enough to worry about with foreign cars and electronics and international cash flow, Japan and Europe also have energetic space programs. But does exploration have to mean "me first" or "we're number one"?

The fourth concept for exploration, the one that seems so obvious in a democratic system that it escapes notice, is that exploration is a choice. Exploration is a social or cultural enterprise that competes for funding and support. It is not guaranteed. It is a matter of priority, weighed in the balance with other programs to satisfy other needs. Exploration occurs when a value judgment is made that it is worth doing and affordable. In our democracy, this decision must be made by consensus of the executive and legislative branches of government and the public at large. Today, we are keenly aware of the many wor-

* Sally K. Ride, *Leadership and America's Future in Space: A Report to the Administrator* (August 1987); Thomas P. Stafford, *America at the Threshold: Report of the Synthesis Group on America's Space Exploration Initiative* (Washington, D.C.: Government Printing Office, 1991).

thy claims on limited fiscal resources, but we are having a difficult time making choices. We want to have it all and do it all, or we make fatalistic claims that doing one completely rules out the other, instead of finding creative ways to do both frugally.

MAKING THE CHOICE

LET US consider a classic pattern for the enterprise of exploration. Someone—a Christopher Columbus or Wernher von Braun, a committee of scientists or presidential advisers—desires to explore some uncharted territory, to discover what is there, to make history, to demonstrate technological prowess or satisfy some other motive. The next step is to gather the necessary

resources: materials, personnel, and funding. Unless you are a king or despot, you must persuade others who control the purse strings to support the effort. To persuade them, you subject the idea to peer review, a process of rigorous critique to ensure that the proposal is sound and justifiable. Columbus spent more than a decade frustrated at this stage; concepts for a manned mission to Mars have been held up at this point for almost 50 years! Winning the endorsement of peers, you move to the next rung on the ladder, the policy makers who decide what is worth doing. You prepare briefing packages to present the reasons and anticipated benefits of exploration. You muster support, give testimony, and drum up public interest. If the policy makers are convinced, the next step is to appeal for funding. Rarely do you get any money the first time around, and even more rarely do you get as much as you think you need. In your enthusiasm, you may exaggerate the anticipated benefits or underestimate the cost,

In words and imagery evoking the exploration of the American West, space is called the new or next or high frontier, and it is assumed that our destiny is to explore there.
NASA

risks, and effort involved, and you probably will have to make periodic appeals for more support.

What does it take to win approval for exploration? A perceived benefit or favorable return on the investment. Affordability. Compatibility with other important motives, such as political priorities. A sense of challenge or mission. And some intangible appeal and inspiration. In the words of a recent report on our future in space, "perhaps the most important space benefit of all is intangible—the uplifting of spirits and human pride in response to truly great accomplishments."*

Ferdinand and Isabella finally approved Columbus's enterprise of the Indies—after declining support for years on the advice of the review committees—when they defeated the Moors and were no longer preoccupied with war, when they realized that the initial cost was trivial compared with the wealth to be gained if the mission succeeded, and when they feared that a neighboring king might beat them to it. The Apollo program to the moon won speedy approval because the perceived Soviet threat in space and in the international ideological struggle was deemed urgent; space exploration was judged to be essential to national security. The requisite consensus for a human mission to Mars has not yet been achieved.

---

* *Report of the Advisory Committee on the Future of the U.S. Space Program* (Washington, D.C.: Government Printing Office, 1990), 3.

Recent studies and position papers urge the link between leadership, competitiveness, and space exploration. On the threshold of a new century, what are America's future goals and priorities in space?

NASA

Sometimes approval does not come, or it is withdrawn when priorities change. A century before Columbus crossed the ocean toward Asia, the Chinese sailed south and west around India and along the eastern coast of Africa. Had they rounded the Cape of Good Hope and sailed north to Europe, the annals of world history would doubtless be much different. Instead, Chinese seafaring exploration ceased with a change in the ruling party at home. Attention turned inward to domestic concerns: agriculture, roads, and the like. Similarly, after six brief landings on the moon, the rest of the Apollo program was canceled; no more Saturn launch vehicles were produced, and no follow-on program for a lunar base or Mars mission was approved. A few more robots were dispatched to the planets, but the United States almost quit exploring space after 1975.

Part of the difficulty in gaining support and approval for exploration is that the results cannot be predicted with certainty. Since exploration takes us into the unknown, we cannot really know what we will find and how worthwhile the discoveries will be until after the fact. Columbus sought Asia but stumbled on the Americas, a world vaster and richer than anyone in Spain could have predicted. The consequences of his voyages were unanticipated: the immediate surge of expeditions led to transoceanic migration and an unprecedented exchange of people, plants, animals, customs, and microbes that changed the world; a flood of wealth into Europe and subsequent rampant inflation; riches gained not from the silk and spice trade, but from gold, silver, sugar, and tobac-

co; the spread of Christianity, accompanied by the subjugation of indigenous peoples and destruction of their cultures; and revolutions in thought about human nature, theology, and knowledge that blossomed in the European Renaissance, Reformation, and Enlightenment.

The immediate results of the lunar-exploration program were better foreseen, although it is still too early to evaluate the long-term results. The mobilization of funds, manpower, and popular consensus to meet a focused goal on a tight schedule in full view of the world produced a triumph that was both technological and political. At the outset, we hardly knew how to reach the moon, much less send people there and return them safely. In less than a decade, the engineering and science were done, the infrastructure was built, the government-industry-research partnership was working effectively, the spin-

offs and practical benefits were appearing, and a dozen men walked on the moon. The success of American democracy and free enterprise amazed the world. And people the world over experienced a triumph of the human spirit and a new sense of the heroic. For about a decade, the lunar program was the apotheosis of exploration as conquest, destiny, and leadership.

Then what happened? The motives for exploration proved to be transient, and the United States shifted gears from human missions to less expensive robotic explorers in the 1970s. The last person left the moon in December 1972. About one-third of the world's population has been born since then. If you are younger than 20, lunar exploration ended before you were born. In your lifetime, we have hardly strayed from Earth except to spend some time in low orbit skimming the atmosphere. Since the last lunar mission, the United States has sent eight successful robotic explorers to fly past all the planets except Pluto, to take a closer look at Venus and Mars, and to land on Mars. We pioneered the exploration of Mars with the *Mariner* spacecraft in the 1960s and 1970s and the *Viking* landers in 1976; but not until 1993 did we attempt to return, this time with an orbiting observatory (which unfortunately did not succeed), and there are no concrete plans for another lander. The *Galileo* spacecraft is going to Jupiter for a closer look, the first since *Voyager* passed by in 1982. Until the *Magellan* orbiter began its extraordinary mapping mission at Venus in 1991, the United States had not sent a spacecraft there since 1978,

The most the United States ever spent on space exploration, in the mid-1960s, was about 4.5 percent of the federal budget. Since then, funding has fallen and stabilized at a level of 1 to 2 percent of the nation's budget.

National Air and Space Museum

although the Soviets kept sending landers (six successful ones) and orbiters to Venus until 1985.

There is popular, grass-roots enthusiasm for space exploration, and presidents periodically announce exploration initiatives with a flourish. However, as a nation we seem to be treating exploration as a low priority. The most the United States ever spent on space exploration, in the mid-1960s, was about 4.5 percent of the federal budget, less than 1 percent of the gross national product. Since then, funding has fallen and stabilized at a level of 1 to 2 percent of the federal budget. Our ambivalence about goals, methods, and priorities has been widely noted in the press and recently by the Advisory Committee on the Future of the U.S. Space Program (Augustine report): among a number of concerns, the first noted is "the lack of a national consensus as to what should be the goals of the civil space program and how they should in fact be accomplished."*

For decades, Mars has been the next target for exploration, but we are still at least 35 million miles and 20 years away from going there. Mars is near enough and enough like Earth that human missions are technically feasible. The problem is not technical; it is choice.

Since the 1950s, would-be explorers have been making the case for a mis-

* *Report of the Advisory Committee on the Future of the U.S. Space Program*, 2.

sion to Mars. In 1969, the Space Task Group reported that "NASA has the demonstrated organizational competence and technology base, by virtue of the Apollo success and other achievements, to carry out a successful program to land man on Mars within 15 years" (that is, by 1985!).* In 1989, the Ride report on leadership recommended a human mission to Mars within the first decade of the twenty-first century: "This leadership initiative . . . would clearly rekindle the national pride and prestige enjoyed by the U.S. during the *Apollo* era. Humans to Mars would be a great national adventure; as such, it would require a concentrated massive national commitment . . . for many decades."† Speaking at the National Air and Space Museum in 1989 on the twentieth anniversary of the first *Apollo* landing, President George Bush urged a "journey into tomorrow, a manned mission to Mars." The Augustine report in 1990 acknowledged that "the long term magnet for the manned space program is the planet Mars—the human exploration of Mars. . . . [S]uch an undertaking probably must be justified largely on the basis of intangibles."‡ Neither the intangibles nor the tangibles—the reasons and costs and benefits—have been articulated well enough to build the consensus necessary for approval. The idea is still mired in the first steps of promotion.

This indecision reflects our ambivalence about the motives for exploration

* *Post-Apollo Space Program*, ii. † Ride, *Leadership and America's Future in Space*, 32–33.
‡ *Report of the Advisory Committee on the Future of the U.S. Space Program*, 6.

and our difficulty in determining whether it is worth it. Why haven't we decided to do something that is technically within our range?

The key is sustained commitment. Exploration is a big enterprise. It takes government support, corporate investment, a large and dedicated team, much preparation and attention to logistical details, and leadership—all of which must be sustained over a period of years, even decades. This kind of commitment is difficult to achieve in a world that operates on short-term interests. Even Columbus could not do it alone; he repeatedly appeared at court to appeal for support. Lewis and Clark carried a letter of credit from President Jefferson in case they got in a pinch. And NASA, like Columbus, makes annual appeals for funding to keep the enterprise of exploration alive.

Another factor is payoff. How do the costs of exploration compare with the benefits? We want a good return on our investment. Spain never dreamed of the wealth and power that came from its modest subsidy of Columbus. It has been estimated that every dollar invested in space-program activities produces $7 to $9 elsewhere in the economy. We can barely imagine the nonmonetary benefits of exploration—for example, how it would change our science and belief to discover life on Mars. Exploration takes us into the unknown, where we cannot predict what we will find, learn, discover, gain. But we are hesitant to pay for something when we are not assured that the payback will be high.

Since the 1950s, proponents have been making the case for a human mission to Mars, but no one has decided to do it.

Paul Hudson, *Morning on Mars* (1992)

A final factor is risk. Exploration is risky business, whether the hazards are shipwreck, bears in the woods, equipment failure, or solar-flare radiation in space. Explorers accept it and try to minimize it, but some still die. As taxpayers whose support is necessary for space exploration, we are cautious about accepting too much risk. We want to be sure there will be no accidents and no loss of life. It is difficult to get that assurance.

FACING THE CHALLENGES

IF WE CHOOSE to go to the planets, we recognize that it is a long, expensive trip, with some hazard to human health and safety. But are these conditions prohibitive? Probably not, but we do face two formidable challenges in planetary travel: radiation and microgravity.

Radiation exposure is the more daunting hazard. Beyond the protective atmosphere and magnetic field of Earth, there is no natural shelter from high-energy radiation. Space explorers will be exposed to dangerous radiation—cosmic rays and solar-flare particles—that may lead to increased risk of cancer or could be lethal. Space is a hostile environment, and there is still much work to be done to understand the radiation risk and to develop effective methods of shielding to reduce the risk to acceptable levels.

Microgravity, or weightlessness, is the other problem. Our bodies are designed to work in gravity. In space, the body changes, adapting to its new environment. Fluids accumulate in the chest and head, and the body's regulatory mechanisms respond. The heart, lungs, blood vessels, kidneys, and glands all react to weightlessness. Gradual changes also begin in the muscles and bones, which become weaker as they no longer feel the load of gravity.

Some of these changes are healthy and reversible; others may not be. How do we keep crew members on long missions—up to 3 years for a round trip to Mars—healthy and fit? After spending such a long time in space, will they be able to function normally after their return to Earth? There is much work to be done to understand how the body changes in microgravity and how those changes can be mitigated by exercise and other measures.

A solution to the problems of radiation and microgravity may be to make space travel faster. If travelers spend less time in space, they should be less vulnerable. Right now, we are using the simplest but most constrained type of propulsion—chemical rockets. They get a fast start, but then coast the rest of the way after their fuel is used up. For decades, we have considered solar- and nuclear-powered propulsion systems, but none has been approved for development—largely for political reasons. There are faster and probably cheaper ways to go to the planets, but advanced transportation systems are not being developed.

Getting explorers to other planets safely, in good health, and quickly are

There are faster, possibly cheaper alternatives to conventional rocket propulsion, but there is no consensus yet to support the development of such vehicles to take human explorers to neighboring worlds.

Pat Rawlings, *The Circuit* (1991)

challenges, but probably not insurmountable ones. We must respect the serious radiation hazard and acknowledge crew health concerns. However, interplanetary travel—at least to the nearby world Mars—is possible and feasible. So why are we not doing it?

Apart from the issues mentioned earlier—commitment and priorities—there is one other big debate: whether people or robots make better explorers. This debate centers on their respective abilities, costs, and risks.

Robotic spacecraft have conducted the preliminary reconnaissance of the solar system with stunning success. We owe most of our knowledge of other worlds to machines that are excellent information collectors. They are lightweight, are easier and cheaper to maintain than people, and can work for years without tiring, although they are not failure-proof. We use them for global mapping and imaging, atmospheric analysis, weather monitoring, soil analysis, and tedious or hazardous work.

The *Viking* robotic probe of Mars, from 1976 to 1982, gave us a wealth of information about this nearby world. Two orbiters surveyed the red planet, and two landers examined the soil on the plains for signs of life. It was thought that *Viking* would lay the foundation for a human mission to Mars. However, the next mission to Mars (16 years later!) was another robot, the *Mars Observer*, launched in 1992 to do a follow-up study of the planet. Plans for a series of landers and sample return missions fell to budget cuts, and talk of a human

mission remains just that—talk. It has been estimated that a robot can visit Mars for one-tenth the cost of a human explorer.

But is cost the only, or even the most important, factor? As explorers, people have the many advantages of innate intelligence. We reason, interpret, and exploit prior experience in response to the unknown and unforeseen. We plan, solve problems, and make decisions. People are well suited to explore and discover on the frontiers of knowledge. However, it is difficult for people to exist on the frontiers of space, because we have to take all our essentials with us. Nevertheless, we apparently cannot resist venturing into the unknown, even when it seems to forbid our presence. The precedents for planetary exploration can be found in human forays into the ocean depths, the frigid Antarctic, and the airless moon.

The division of labor between people and machines is constantly changing. It is becoming possible to "feel" present in another environment by sensory feedback, using techniques of telepresence and virtual reality. These machines provide such realistic impressions to remote-control operators that they have the illusion of being in the computer-generated environments themselves. Although artificial hands and brains are not yet as agile as a human's, these new techniques may blend human intelligence and robotic operations. What will happen then? Will people need to go into space to explore? Will they still want to?

Visions of people living and working on Mars tantalized early space pio-

There is considerable debate over the comparative role of people and robots in space exploration. Each has capabilities and risks, but debate tends to focus on differences in cost.

Overleaf: Pat Rawlings, *Return to Utopia* (1991)

neers and excite many futurists today. To live on Mars, people must bring or create their own air, water, food, and shelter. They may have to live underground or in habitats covered with soil, because the feeble atmosphere there does not provide enough protection against radiation exposure. The first explorers may camp out in landers that descend from an orbiting mission spacecraft for a short stay of a few weeks and then launch to rejoin the orbiter for a return voyage to Earth. For longer stays, they might build permanent structures on the surface or underground. Residents would use local resources to supply some of their basic needs.

Explorers on Mars probably would go on expeditions for geologic research or other remote tasks. They will need vehicles and protective suits similar to those used on the moon. For short errands, an open-cab "dune buggy" might suffice. For longer trips, they might prefer something more like a Winnebago, a home on wheels.

People will change the surface of whatever worlds they visit. If Mars is a target for eventual settlement, space explorers would be wise to learn some lessons from our stewardship of our first planetary home.

People dream of exploring beyond the solar system, but the difficult challenges of interplanetary travel are trivial compared with those of voyaging to the stars. Cosmic distances and travel times are so great that conventional notions of exploring are hardly relevant.

Pat Rawlings

If the goal of interstellar exploration is to increase knowledge, there is a more efficient method of collecting information than travel—the use of optical and radio telescopes. It is not always necessary to go somewhere to explore. We can search for evidence of other worlds around other stars by using telescopes on Earth and in space. Without leaving home, we can listen for radio signals from other civilizations through SETI (Search for Extraterrestrial Intelligence).

If direct contact with other worlds is the motive, however, we will need new technologies and a strong will to go to the stars. Interstellar travel is about as feasible now as space flight was in the era of Columbus. It is not possible with today's technology and budgets or any foreseeable in the near future, although technical breakthroughs could change that. Many kinds of futuristic spacecraft and propulsion systems—antimatter, fusion, ramjets—have been envisioned. Only 66 years, a person's lifetime, elapsed between the first powered airplane flight and the first manned landing on the moon. In less than one generation, we developed the means for orbital and planetary space flight. Might interstellar flight likewise become possible?

BACK TO THE MUSEUM

VISITORS to the National Air and Space Museum increasingly make up a generation born after the *Apollo* landings on the moon, after the

Viking explorations of Mars, and after the *Voyager 1* and *Voyager 2* grand tours of the outer planets. Children have no direct memory of these triumphs. They take the pretty pictures for granted, but have little sense of the effort and enterprise, the challenges and commitments required to venture to the moon and beyond. It is as though the explorations never occurred, so remote are they from our young visitors' experience. The artifacts we display in silent testimony may seem as ancient as dinosaurs and mummies to children who stroll by on their school field trips and family vacations. At the same time, this generation is entertained by "Star Trek," "Star Wars," and other fictions of a space-faring future that is by no means certain.

I wonder if children notice the gap between futuristic fiction and the aging artifacts of space exploration. When they see the lunar lander or the *Viking* and *Voyager* spacecraft, maybe they ask their parents and teachers, "Did that really happen? Why didn't we stay? Will we go back?" I hope that they realize that the question "Where next?" is theirs—and ours—to answer. If we believe that we are a nation of explorers, we need to make some decisions about our future. Exploration may or may not be in our nature, our destiny, or our character. But it certainly is within our field of choice.

Suggestions for Further Reading

LEGACY: PERSPECTIVES ON PAST EXPLORATION

Armstrong, Neil. *First on the Moon: A Voyage with Neil Armstrong, Michael Collins, and Edwin E. Aldrin, Jr.* Boston: Little, Brown, 1970.

Boorstin, Daniel J. *The Discoverers*. New York: Vintage Books, 1985.

Collins, Michael. *Carrying the Fire: An Astronaut's Journeys*. New York: Farrar, Straus & Giroux, 1989.

Collins, Michael. *Liftoff: The Story of America's Adventure in Space*. New York: Grove Press, 1988.

French, Bevan M., and others. *A Meeting with the Universe: Science Discoveries from the Space Program*. Washington, D.C.: National Aeronautics and Space Administration, 1981.

Goetzmann, William H. *New Lands, New Men: America and the Second Great Age of Discovery*. New York: Viking, 1986.

Hale, J. R. *Age of Exploration*. New York: Time, 1966.

Hale, J. R. *Renaissance Exploration*. New York: Norton, 1972.

Logsdon, John H. *The Decision to Go to the Moon: Project Apollo and the National Interest*. Cambridge, Mass.: MIT Press, 1970.

McDougall, Walter A. *The Heavens and the Earth: A Political History of the Space Age*. New York: Basic Books, 1985.

Michener, James, and others. *Why Man Explores*. Washington, D.C.: National Aeronautics and Space Administration, 1976.

Parry, J. H. *The Age of Reconnaissance*. New York: New American Library, 1963.

Parry, J. H. *The Discovery of the Sea*. Berkeley: University of California Press, 1981.

Pyne, Stephen J. *The Ice: A Journey to Antarctica*. Iowa City: University of Iowa Press, 1986.

Voyager, the Grandest Tour: The Mission to the Outer Planets. Pasadena, Calif.: National Aeronautics and Space Administration, 1991.

CHALLENGE: VIEWS ON EXPLORATION TODAY

Boorstin, Daniel J. *The Exploring Spirit: America and the World, Then and Now*. New York: Random House, 1976.

Bradbury, Ray, and others. *Mars and the Mind of Man*. New York: Harper & Row, 1973.

Ferris, Timothy. *Coming of Age in the Milky Way*. New York: Morrow, 1988.

Gould, Stephen Jay. *Wonderful Life: The Burgess Shale and the Nature of History*. New York: Norton, 1989.

Hargrove, Eugene C. *Beyond Spaceship Earth: Environmental Ethics and the Solar System*. San Francisco: Sierra Club Books, 1986.
Murray, Bruce C. *Journey into Space: The First Three Decades of Space Exploration*. New York: Norton, 1990.
Oliver, Jack E. *The Incomplete Guide to the Art of Discovery*. New York: Columbia University Press, 1991.
O'Neill, Gerard K. *The High Frontier: Human Colonies in Space*. New York: Morrow, 1977.
Sagan, Carl. *Broca's Brain: Reflections on the Romance of Science*. New York: Random House, 1979.
Sagan, Carl. *The Cosmic Connection: An Extraterrestrial Perspective*. Garden City, N.Y.: Anchor Books, 1973.
White, Frank. *The Overview Effect: Space Exploration and Human Evolution*. Boston: Houghton Mifflin, 1987.

DESTINY: QUESTIONS FOR FUTURE EXPLORERS

Exploring the Moon and Mars: Choices for the Nation. Washington, D.C.: Congress of the United States, Office of Technology Assessment, 1991.
Finney, Ben R., and Eric M. Jones. *Interstellar Migration and the Human Experience*. Berkeley: University of California Press, 1985.
Forward, Robert L. *Future Magic*. New York: Avon Books, 1988.
Forward, Robert L., and Joel Davis. *Mirror Matter: Pioneering Antimatter Physics*. New York: Wiley, 1988.
Jastrow, Robert. *Journey to the Stars: Space Exploration, Tomorrow and Beyond*. New York: Bantam Books, 1989.
Mallove, Eugene F., and Gregory L. Matloff. *The Starflight Handbook: A Pioneer's Guide to Interstellar Travel*. New York: Wiley, 1989.
Pioneering the Space Frontier: The Report of the National Commission on Space. New York: Bantam Books, 1986.
Sagan, Carl. *Cosmos*. New York: Random House, 1980.

Index